Lecture Notes in Control and Information Sciences

Edited by M. Thoma and A. Wyner

138

D. C. McFarlane, K. Glover

Robust Controller Design Using Normalized Coprime Factor Plant Descriptions

Springer-Verlag
Berlin Heidelberg New York
London Paris Tokyo Hong Kong

Authors
Duncan C. McFarlane
BHP Melbourne Research Laboratories
P.O. Box 264
Clayton 3168
Victoria
Australia

Keith Glover
Dept. of Engineering
University of Cambridge
Trumpington Street
Cambridge CB2 1PZ, UK

ISBN 3-540-51851-7 Springer-Verlag Berlin Heidelberg New York
ISBN 0-387-51851-7 Springer-Verlag New York Berlin Heidelberg

Library of Congress Cataloging-in-Publication Data
Glover, K. (Keith),
Robust controller design using normalized coprime factor plant design /
K. Glover, D. C. McFarlane.
(Lecture notes in control and information sciences ; 138)
Includes bibliographical references.
ISBN 0-387-51851-7 (U.S. : alk. paper)
1. Automatic control. 2. Control theory.
I. McFarlane, D. C. (Duncan C.). II. Title. III. Series.
TJ213.G525 1989
629.8–dc20 89-21982

Printed in Germany

Printing: Mercedes-Druck, Berlin
Binding: B. Helm, Berlin
2161/3020-543210 Printed on acid-free paper.

Acknowledgements

Duncan McFarlane would like to thank the B.H.P. Co. Ltd., Australia, for their support of this work, which was performed while he was a Ph.D. student at the University of Cambridge.

The authors are pleased to acknowedge the following: M. Vidyasagar for valuable discussions in the early stages of this work; B. Francis, N.J. Young, R.J. Ober, J.R. Partington, D. Mustafa and P.I. Iglesias for helpful comments on the manuscript; J.M. Maciejowski and M. Noton for their cooperation with the design examples; and E. Donaldson for her assistance in preparing the manuscript.

TABLE OF CONTENTS

LIST OF FIGURES

NOTATION

Symbols

Note: Symbols in boldface are functions of the Laplace Transform variable 's'. All transfer functions matrices are taken to be matrices with real-rational elements only, and constant matrices are taken to be matrices with real elements only.

$\mathcal{C}$	field of complex numbers
$\Re$	field of real numbers
$j\Re$	the imaginary axis of the complex plane
R	see Definition 2.9
RL_2	see Definition 2.10
RH_2	see Definition 2.10
RL_∞	see Definition 2.11
RH_∞	see Definition 2.11
$\lambda_i(\bullet)$	the ith eigenvalue
$\rho(\bullet)$	$=\lambda_{max}(\bullet)$ – spectral radius
$\sigma_i(\bullet)$	the ith singular value
$\overline{\sigma}(\bullet)$	the maximum singular value
$\underline{\sigma}(\bullet)$	the minimum singular value
$c(\bullet)$	$= \overline{\sigma}(\bullet)/\underline{\sigma}(\bullet)$ – the condition number
$\mu_i(\bullet)$	the ith characteristic value – see Theorem 4.26
σ_i	the ith Hankel singular value – see Definition 2.2
$\\| \bullet \\|_2$	L_2/H_2 norm – see Definition 2.10
$\\| \bullet \\|_\infty$	L_∞/H_∞ norm – see Definition 2.11
$\vert \bullet \vert$	modulus
$\\| \bullet \\|_H$	Hankel Norm – see Definition 2.2
$A_\perp$	orthogonal complement
$A > 0$	A positive definite
$A > B$	$A - B$ is a positive definite matrix
$A \geq 0$	A positive semi-definite
A^*	complex conjugate transpose
I_n	$n \times n$ identity matrix
$\mathbf{G}^*$	$= G(-\bar{s})^*$ – complex conjugate transpose
$Re(\bullet)$	real part
$Ric(\bullet)$	Riccati operator – see Section 3.4.4
$E(\bullet)$	expectation

$\eta(\bullet)$	number of closed right half-plane poles
$\mathcal{F}_L(\boldsymbol{P},\boldsymbol{K})$	Lower Linear Fractional Transformation – see Definition 2.13
$\mathcal{F}_U(\boldsymbol{P},\Delta)$	Upper Linear Fractional Transformation – see Definition 2.13
$\mathcal{T}_U[\boldsymbol{K}]$	Alternative form for a Lower Linear Fractional Transformation – see Lemma 2.14
$\mathcal{H}(\boldsymbol{G},\boldsymbol{K})$	closed-loop transfer matrix – Section 5.2
$\mathcal{D}$	class of uncertainty
$\mathcal{D}_\epsilon, \mathcal{D}_{\mathcal{S}_\epsilon}, \mathcal{D}_{\mathcal{U}_\epsilon}$	classes of bounded uncertainty – see Definition 3.1
(A,B,C,D)	state-space realization – Section 2.2
^{o}	degrees
$\simeq, \lesssim, \gtrsim$	asymptotically equal to, less than or equal to, greater than or equal to – Section 6.4.1

Acronyms

SISO	Single-Input, Single-Output
MIMO	Multi-Input, Multi-Output
LTI	Linear, Time-Invariant
RHP,LHP	Right Half-Plane, Left Half-Plane
RCF	Right Coprime Factorization
LCF	Left Coprime Factorization
LFT	Linear Fractional Transformation
LLFT	Lower Linear Fractional Transformation
ULFT	Upper Linear Fractional Transformation
ARE	Algebraic Riccati Equation
GCARE	Generalized Control Algebraic Riccati Equation
GFARE	Generalized Filter Algebraic Riccati Equation
LTR	Loop Transfer Recovery
LQR	Linear Quadratic Regulator
LQG	Linear Quadratic Gaussian
LSDP	Loop Shaping Design Procedure
BTA	Balanced Truncation Approximation
HNA	Hankel Norm Approximation
S & T	Sensitivity and Complementary Sensitivity

CHAPTER ONE

INTRODUCTION

1.1 Background and Problem Description

The modelling of a physical system for feedback control invariably involves a trade-off between the simplicity of the model and its accuracy in matching the behaviour of the physical system. At best, it is only possible to obtain a model which is, in some sense, *close* to the actual system. This motivates the need for *robustly stabilizing* or *robust* feedback controllers. As well as satisfying the minimum requirement of stabilizing the nominal system model, a robust controller also stabilizes a *class* of systems which are, in some sense, *close* to the nominal plant. Using this approach, if the actual physical system is contained in this class of systems, a robust controller *guarantees* to stabilize it.

There are numerous ways of representing classes of systems that are 'close' to a nominal model, and in this monograph, we adopt the 'coprime factor' framework of Vidyasagar, (1985) to represent both the nominal model transfer function, and a class of 'close' systems: if G is the nominal plant transfer function, then we write

$$G = \tilde{M}^{-1}\tilde{N} \tag{1.1}$$

where $\tilde{N}$, $\tilde{M}$ are stable transfer functions representing a *left coprime factorization* of G. We then define the transfer functions of a set of 'close' systems by

$$G_{\Delta} = (\tilde{M} + \Delta_M)^{-1}(\tilde{N} + \Delta_N) \tag{1.2}$$

where Δ_N, Δ_M are stable, unknown transfer functions satisfying the condition $\|[\Delta_N, \Delta_M]\|_{\infty} < \epsilon$. That is, transfer functions Δ_N, Δ_M, represent unknown variations in G which are limited in size. Note that Δ_N and Δ_M can therefore be thought of as representing uncertainty in the nominal plant model.

The *robust stabilization* problem of finding a single feedback controller which stabilizes all G_Δ in (1.2) for ϵ as large as possible, can be solved using the H_∞ *optimization methods* first introduced by Zames (1981): an H_∞ optimization problem involves finding a controller which (a) stabilizes the closed-loop system and (b) minimizes the 'size' (H_∞ norm) of a particular closed-loop transfer function which represents a design objective. (See Francis, 1987 and the references therein, for details.) Solutions to H_∞ optimization problems are typically iterative in nature. Returning to the robust stabilization problem mentioned above, a stabilizing controller achieving the largest value of ϵ, ϵ_{max}, is also a stabilizing controller which minimizes the H_∞ norm of a related closed-loop transfer function, and can be obtained by solving this H_∞ optimization problem. Such a controller is referred to as an *optimally robust controller* and ϵ_{max} is the *maximum stability margin* for this problem.

Further, if the nominal plant description in (1.1) is chosen so that the factors $\tilde{N}$ and $\tilde{M}$ are *normalized*, (we will clarify this later), then we will show that the optimally robust controller can be obtained without iteration, and an explicit formula for ϵ_{max} can be derived. We call the problem of finding such a controller the *normalized coprime factor robust stabilization problem*, and it provides the basis of the work contained in this monograph.

The design of effective robust controllers for Multi-Input, Multi-Output (MIMO) systems is a difficult task. The 'final product' of this monograph is a methodical approach to the design of robust controllers for MIMO systems using H_∞ optimization methods. This approach incorporates the normalized LCF robust stabilization problem into a three stage design procedure, and provides a control designer with a simple and logical approach to MIMO controller design.

1.2 Organization

This monograph is divided into seven chapters, and a summary of these will now be given.

Chapter 2: Preliminaries

In this chapter we introduce a number of well known results in control theory which are central to later chapters. We introduce the H_∞ and H_2 *norms* as measures of the gain of a transfer function, *linear fractional transformations* are defined, a number of important results from *coprime factorization theory* are stated, and finally we review the subject of closed-loop stability.

Chapter 3: Robust Stabilization of Uncertain Systems

This chapter examines a number of frequency domain approaches to modelling classes of uncertain systems, where all elements in the class are close, in an H_∞ norm sense, to a nominal system model. It is shown that a single standard framework can be used to represent many such classes. Further, using this standard framework, necessary and sufficient conditions are given for a single, linear, time-invariant (LTI) controller to stabilize all systems in a particular uncertainty class, and hence we can pose a *robust stabilization problem.* The coprime factor uncertainty description of (1.2) is shown to have advantages over other common uncertainty descriptions in this context. The problem of robust stabilization is shown to be equivalent to an H_∞ *optimization problem*, and an iterative solution procedure for H_∞ optimization problems, due to Glover and Doyle, (1988), is outlined.

Chapter 4: Robust Stabilization of Normalized Coprime Factor Plant Descriptions

In this chapter we give an explicit solution to the *normalized LCF robust stabilization problem* introduced in Section 1.1 and show that ϵ_{max}, the *maximum stability margin*, can be obtained without iteration. A coprime factorization of all optimal controllers for this problem can be generated directly from the normalized left coprime factorization of the nominal plant G via a related *Nehari extension problem.* The related problem of finding so called *sub-optimal* controllers (controllers achieving a specified level of robust stability which is less than ϵ_{max}) for this problem is also introduced, and a *state-space characterization* of all such controllers is given. We then compare this solution for the optimal controllers to the solution obtained using the general H_∞ procedure of Glover and Doyle, (1988) and show why the iteration, implicit in this procedure, can be avoided. We show that a related class of H_∞ optimization problems also have exact (non-iterative) solutions, and finally, the special class of nominal plants which admit *passive* state-space realizations is examined in the normalized left coprime factorization framework.

Chapter 5: Reduced Order Controller Design

Here we examine the robust stability properties of a closed-loop system when a controller solving the normalized LCF robust stabilization problem is replaced by a reduced order approximant. A *model reduction algorithm*, also based on coprime factor model representations is reviewed, and two procedures for designing *reduced order controllers* are presented, both of which incorporate the above mentioned model reduction algorithm, and the normalized left coprime factorization robust stabilization problem. Robust stability guarantees are shown to exist for reduced order controllers designed using these procedures.

Chapter 6: A Loop Shaping Design Procedure

It is important in controller design to not only achieve *robust stability* properties, but to also achieve a specified level of *performance*. In order to incorporate performance objectives into the robust controller design approach of Chapter 4, we propose a three stage design procedure which uses a *loop shaping* approach to introduce performance/robustness trade-offs, and uses the normalized coprime factor robust stabilization method to guarantee closed-loop stability and a certain level of robust stability. The procedure is then as follows:

(i) Using pre and/or post compensation, the nominal plant singular values are 'shaped', using a simple loop shaping method, to give a desired open-loop shape. A 'shaped' plant is formed by combining the nominal plant and these compensators.

(ii) A feedback controller which robustly stabilizes the normalized LCF of the shaped plant is synthesized.

(iii) The feedback controller and shaping compensators are combined to form the final controller.

We give theoretical justification for adopting this technique, and show that it is a simple and systematic approach to design.

Chapter 7: Design Examples

Controllers are produced for three design problems using the design algorithm of Chapter 6, and incorporating the model reduction methods of Chapter 5 when appropriate. The first two examples are the results of a study performed for British Aerospace P.L.C., U.K., on the applicability of this design algorithm for highly flexible spacecraft control. In the third example, an aircraft control problem, we compare two controllers designed using the design procedure of Chapter 6, with controllers designed using two other design techniques.

As a final comment, the reader should note that because of the diversity of the material covered in this monograph, it is not appropriate to review all related work in a single section. Instead, related literature will be reviewed at appropriate places in the monograph. In particular, the reader is referred to Section 3.1 for background on the topics of robust stabilization and H_∞ optimization, to Section 5.2 for a review of model reduction techniques, and work on the coprime factor model reduction problem, and finally to Section 6.2 for a summary of loop shaping methods.

CHAPTER TWO

PRELIMINARIES

2.1 Introduction

The aim of this chapter is to introduce a number of well known results in control theory which are particularly applicable to the work in the following chapters of this monograph. No new material is presented here, and the results stated can be considered to be purely preliminary in nature. In Section 2.2, standard state-space forms are revised, and results relating to the controllability and observability of a state-space system are given. In addition, the important topics of inner, co-inner, and all-pass systems, which are central to the developments in Chapters 4 and 5, are reviewed. Section 2.3 introduces frequency domain spaces of real-rational transfer functions, and definitions of the H_∞ and H_2 norms are stated. In Section 2.4 we state two representations for a Linear Fractional Transformation (LFT), which provides a standardized framework for a variety of feedback arrangements. The subject of coprime factorizations is introduced in Section 2.5. The results in this section provide a necessary basis for the remainder of this monograph which is concerned, in particular, with *normalized* coprime factorizations, introduced in Definition 2.19. Finally, in Section 2.6 we review some closed-loop stability results, introducing the notion of internal stability as the now accepted condition for closed-loop stability. Incorporating results from Sections 2.4 and 2.5 we give a result (due to Doyle and Chu, 1986) which guarantees internal stability for a very general class of feedback systems. For convenience, the Small Gain Theorem, due to Zames, (1966), and the Generalized Nyquist Criterion, originally due to Rosenbrock, (1972), are also stated.

2.2 State-Space Systems

2.2.1 State-Space Forms

In this monograph, we will assume that all systems are linear, finite-dimensional and time invariant. State-space forms provide a convenient means of describing the dynamical behaviour of a linear system. A continuous time, state-space system

$$\begin{aligned}\dot{x}(t) &= Ax(t) + Bu(t)\\ y(t) &= Cx(t) + Du(t)\end{aligned}$$

will be denoted (A, B, C, D) or $\left[\begin{array}{c|c} A & B \\ \hline C & D \end{array}\right]$ where $A \in \Re^{n\times n}$, $B \in \Re^{n\times m}$,$C \in \Re^{p\times n}$, and $D \in \Re^{p\times m}$. The corresponding $p \times m$ transfer function, with (MacMillan) degree n, is given by

$$G(s) = C(sI - A)^{-1}B + D$$

and we adopt the convention $G(s) \stackrel{s}{=} (A, B, C, D)$ or $G(s) \stackrel{s}{=} \left[\begin{array}{c|c} A & B \\ \hline C & D \end{array}\right]$. Under a state similarity transformation, T, this becomes

$$G \stackrel{s}{=} \left[\begin{array}{c|c} TAT^{-1} & TB \\ \hline CT^{-1} & D \end{array}\right],$$

where G is used as a short form for $G(s)$. Finally, if D is invertible, a state-space representation of G^{-1} is given by

$$G^{-1} \stackrel{s}{=} \left[\begin{array}{c|c} A - BD^{-1}C & BD^{-1} \\ \hline -D^{-1}C & D^{-1} \end{array}\right].$$

2.2.2 Controllability and Observability

It will be assumed in this work that the reader is familiar with the basic concepts of controllability, observability, and minimality as outlined in most standard control textbooks. We will now define some properties of minimal systems. The controllability and observability Gramians are of fundamental importance, and are of particular use in model reduction. (See Chapter 5.)

Definition 2.1

Let G be stable with a minimal state-space realization (A, B, C, D). The associated ***controllability Gramian****, P, and* ***observability Gramian****, Q, are defined as:*

$$P \triangleq \int_0^\infty e^{At} BB^* e^{A^*t} dt$$
$$Q \triangleq \int_0^\infty e^{A^*t} C^*C e^{At} dt.$$

Further, it can be shown that P and Q are also the unique, positive definite solutions to the following Lyapunov equations:

$$AP + PA^* + BB^* = 0 \tag{2.1}$$

$$A^*Q + QA + C^*C = 0 \tag{2.2}$$

The Hankel singular values of G, defined next, are invariants under state-space transformation, and are indicative of the controllability and observability properties of system states in a particular ('balanced') state-space realization. (See Section 5.2.)

Definition 2.2

With the notation in Definition 2.1, the ***Hankel singular values*** *of G with degree n are given by*

$$\sigma_i \triangleq \lambda_i^{1/2}(PQ), \quad i = 1, \ldots, n$$

ordered by convention, $\sigma_1 \geq \sigma_2 \ldots \geq \sigma_n \geq 0$. The ***Hankel Norm****, denoted $\| \bullet \|_H$ is defined to be σ_1.*

Related to controllability and observability are the concepts of stabilizability and detectability. Consider a (not necessarily minimal) state-space realization (A, B, C, D):

Definition 2.3

The pair (A, B) is ***stabilizable*** *if and only if all unstable modes of (A, B, C, D) are controllable.*

Definition 2.4

The pair (C, A) is ***detectable*** *if and only if all unstable modes of (A, B, C, D) are observable.*

2.2.3 Inner and Co-Inner Systems

Inner and co-inner systems are relevant to many aspects of control theory: For example, Doyle and Chu, (1986), show the importance of inner/co-inner systems in H_∞ optimization, and Glover, (1984) has illustrated the importance of such systems in model reduction. We now define inner and co-inner systems, and some associated properties are given.

Definition 2.5

A stable, $p \times m$, transfer function G, $p \geq m$, is called ***inner*** *if*

$$G^*G = I$$

for all $s \in j\Re$.

Proposition 2.6 (Doyle, 1984)

Let G be a stable, $p \times m$ transfer function, $p \geq m$, with minimal state-space realization (A, B, C, D) and let $Q = Q^* > 0$ be such that

$$A^*Q + QA + C^*C = 0.$$

Then G is inner if and only if

(i) $D^*C + B^*Q = 0$

(ii) $D^*D = I$.

Similar results exist for co-inner systems, which are also stated here for completeness:

Definition 2.7

A stable, $p \times m$, transfer function G, $p \leq m$, is called ***co-inner*** *if*

$$GG^* = I$$

for all $s \in j\Re$.

Proposition 2.8 (Doyle, 1984)

Let G be a stable, $p \times m$ transfer function, $p \leq m$, and let $P = P^* > 0$ be such that

$$AP + PA^* + BB^* = 0.$$

Then G is co-inner if and only if

(i) $DB^* + CP = 0$

(ii) $DD^* = I.$

Finally, note that a square, not necessarily stable transfer function G is called ***all-pass*** if $GG^* = I$ (or equivalently if $G^*G = I$) for all $s \in j\Re$.

2.3 Frequency Domain Spaces

In this section we will briefly discuss frequency domain spaces of real-rational, matrix valued, transfer functions. It should be noted that this discussion is not intended to be comprehensive, and the reader is referred to Francis, (1987) or Rudin, (1986) for further details. Some preliminary definitions are now given.

Definition 2.9

R denotes the space of all real-rational transfer function matrices.

Definition 2.10

The L_2/H_2 norm of G is given by

$$\|G\|_2 \triangleq \sqrt{\frac{1}{2\pi}\int_{-\infty}^{\infty} Trace\,(G^*(j\omega)G(j\omega))\,d\omega}.$$

(a) RL_2 denotes the space of all real-rational transfer function matrices with no poles on the imaginary axis and with finite L_2 norm.

(b) RH_2 denotes all transfer function matrices in RL_2 which have no poles in $Re(s) > 0$.

Definition 2.11

The L_∞/H_∞ norm of G is given by

$$\|G\|_\infty \triangleq \sup_{\omega\in\Re} \overline{\sigma}[G(j\omega)].$$

(a) RL_∞ denotes the space of all real-rational transfer function matrices which have no poles on the imaginary axis.(i.e. with finite L_∞ norm)

(b) RH_∞ denotes all transfer function matrices in RL_∞ which have no poles in $Re(s) > 0$.

The spaces RL_2 and RH_2 are often specialized to the case of vector-valued transfer functions, and for the case of standard control system signals, RH_2 can be shown

(Francis, 1987, Theorem 1) to be the set of Laplace transforms of signals with bounded energy for $t \geq 0$. RH_2 is *isomorphic* to the space of such signals, and the corresponding H_2 norm is the energy contained in the signal. Also note that RH_∞ is simply the space of all asymptotically stable, real-rational transfer function matrices.

The next theorem states an important result in H_∞ optimization theory, relating the H_∞ norm of a transfer function matrix to the H_2 norms of its associated input and output transfer function vectors.

Theorem 2.12

If $G \in RH_\infty$, then $GRH_2 \subset RH_2$ and

$$\|G\|_\infty = sup\{\|\boldsymbol{Gu}\|_2 : \boldsymbol{u} \in RH_2, \|\boldsymbol{u}\|_2 = 1\}.$$

If $\boldsymbol{y} = \boldsymbol{Gu}$, then Theorem 2.12 states that for any input, $\boldsymbol{u}$, of unit energy, the energy in $\boldsymbol{y}$ for $t \geq 0$ is bounded by $\|G\|_\infty$, the H_∞ norm of $\boldsymbol{G}$. $\|G\|_\infty$ is called the *induced norm* of the mapping $\boldsymbol{u} \mapsto \boldsymbol{Gu}$ ($\boldsymbol{u} \in RH_2$, $\|\boldsymbol{u}\|_2 = 1$).

2.4 Linear Fractional Transformations

Linear fractional transformations are frequently used in H_∞ optimization, as well as many other areas of control theory, and they represent a means of standardizing a wide variety of feedback arrangements. A number of basic concepts are outlined here.

Definition 2.13

(a) A (lower) linear fractional transformation (LLFT) is denoted: (See Figure 2.1)

$$\mathcal{F}_L(\left[\begin{array}{c|c} \boldsymbol{P}_{11} & \boldsymbol{P}_{12} \\ \hline \boldsymbol{P}_{21} & \boldsymbol{P}_{22} \end{array}\right], \boldsymbol{K}) \triangleq \boldsymbol{P}_{11} + \boldsymbol{P}_{12}\boldsymbol{K}(I - \boldsymbol{P}_{22}\boldsymbol{K})^{-1}\boldsymbol{P}_{21} \tag{2.3}$$

whenever $det(I - \boldsymbol{P}_{22}\boldsymbol{K}) \neq 0$.

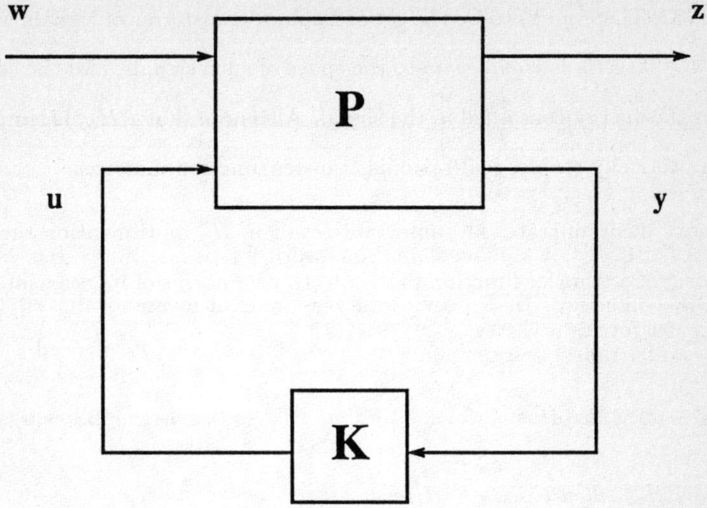

Figure 2.1 Lower Linear Fractional Transformation (LLFT)

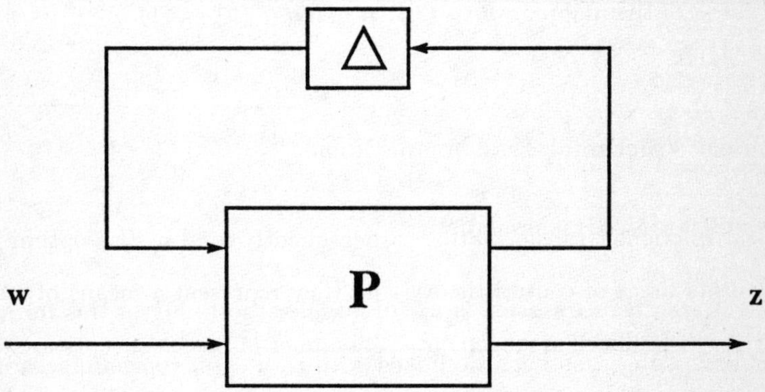

Figure 2.2 Upper Linear Fractional Transformation (ULFT)

(b) An (upper) linear fractional transformation (ULFT) is denoted: (See Figure 2.2)

$$\mathcal{F}_U(\left[\begin{array}{c|c} P_{11} & P_{12} \\ \hline P_{21} & P_{22} \end{array}\right], \Delta) \triangleq P_{22} + P_{21}\Delta(I - P_{11}\Delta)^{-1}P_{12} \tag{2.4}$$

whenever $det(I - P_{11}\Delta) \neq 0$.

The LLFT provides a convenient framework for posing many H_∞ problems (see Section 3.5), while the ULFT provides a very general means of describing uncertain systems. (See Section 3.3 in particular.)

An alternative expression for a LLFT, if P_{21}^{-1} exists, is the following:

Lemma 2.14

Any LLFT in Definition 2.13, such that $det(P_{21}) \neq 0$, *can also be written:*

$$\mathcal{T}_U[K] = (U_{11}K + U_{12})(U_{21}K + U_{22})^{-1} \tag{2.5}$$

where

$$U = \left[\begin{array}{c|c} U_{11} & U_{12} \\ \hline U_{21} & U_{22} \end{array}\right] = \left[\begin{array}{c|c} P_{12} - P_{11}P_{21}^{-1}P_{22} & P_{11}P_{21}^{-1} \\ \hline -P_{21}P_{22} & P_{21}^{-1} \end{array}\right]$$

with $det(U_{21}K + U_{22}) \neq 0$.

Proof:

By simple matrix manipulation. ■

This alternative expression is useful because the LLFT in this form is written as a stable factorization, and can be linked with results on coprime factorizations which are examined next.

The reader is referred to Dym, (1988), Redheffer, (1960) and Ball and Ran, (1986) for further details on LFTs.

2.5 Coprime Factorizations

Coprime factorizations are central to many aspects of control theory, and in particular, many of the fundamental ideas in H_∞ optimization have their basis in coprime factor theory. (See Chu and Doyle, 1986, or Vidyasagar, 1985.) In this work we will use coprime factor descriptions to represent both systems to be controlled and their corresponding controllers. In Chapters 3,4 and 5 we show that this form of description has particular advantages in robust stabilization and model reduction problems.

In this section we formally define coprime factors, and outline some important properties associated with coprime factorizations. The results given here have been restricted to the realm of real-rational and proper transfer function matrices and their associated state-space systems. It is noted that many results on coprime factors can be applied to more general settings. (See Vidyasagar, 1985.) Further, only properties of *left* coprime factorizations are stated here. It should be noted that analogous properties associated with *right* coprime factorizations hold without exception.

2.5.1 Definitions

Firstly, we define a necessary and sufficient condition for two matrices to be left coprime:

Definition 2.15

*Suppose $\tilde{M}, \tilde{N} \in RH_\infty$ have the same number of rows. Then $\tilde{M}$ and $\tilde{N}$ are **left coprime** if and only if there exist U, $V \in RH_\infty$ such that*

$$\tilde{M}V - \tilde{N}U = I. \tag{2.6}$$

It is possible to represent *any* real-rational, proper transfer function in terms of a pair of asymptotically stable, real-rational, proper transfer functions which are left coprime. This is called a left coprime factorization, and is defined next.

Definition 2.16

The pair $(\tilde{N}, \tilde{M})$, where $\tilde{M}, \tilde{N} \in RH_\infty$, constitutes a ***Left Coprime Factorization (LCF)*** *of $G \in R$ if and only if*

(a) $\tilde{M}$ is square, $det(\tilde{M}) \neq 0$

(b) $$G = \tilde{M}^{-1}\tilde{N} \tag{2.7}$$

(c) $\tilde{N}$ and $\tilde{M}$ are left coprime.

There are a number of advantages in representing a possibly unstable transfer function in terms of two stable factors in this way and this will be exploited in later chapters. To motivate the selection of stable, *coprime* factors to represent a transfer function, we note by Vidyasagar, (1985), that in the scalar case such a selection precludes the possibility of the stable factors having common zeros in the closed RHP. Hence the coprime factor representation of a transfer function will contain *no* unstable, hidden plant modes – an essential prerequisite for feedback control design.

We next examine the uniquness properties of the left coprime factorizations that can be generated for a single transfer function $G \in R$. These will be characterized in Lemma 2.18. We firstly define a 'unit' in RH_∞:

Definition 2.17

Any square, invertible, transfer function matrix satisfying U, $U^{-1} \in RH_\infty$ is called a unit in RH_∞.

Lemma 2.18 (Vidyasagar, 1985)

The left coprime factors $\tilde{N}$ and $\tilde{M}$ of a transfer function $G(s) \in R$ are unique to within left multiplication by a unit in RH_∞.

The above Lemma states that there is an arbitrarily large number of coprime factorizations for any one transfer function.

Of particular interest in this monograph is a left coprime factorization of G in which the factors $\tilde{N}$, $\tilde{M}$ are *normalized:*

Definition 2.19

The pair $(\tilde{N}, \tilde{M})$, *where* $\tilde{N}$, $\tilde{M} \in RH_\infty$ *constitute a* ***normalized left coprime factorization*** *of* $G \in R$ *if and only if* $(\tilde{N}, \tilde{M})$ *is a LCF of* G, *and in addition*

$$\tilde{N}\tilde{N}^* + \tilde{M}\tilde{M}^* = I \qquad \text{for all } s \in j\Re. \tag{2.8}$$

Note that (2.8) states that the matrix $[\tilde{N}, \tilde{M}]$ is co-inner.

The next Lemma summarizes the existence and uniqueness properties of a normalized left coprime factorization.

Proposition 2.20 (Vidyasagar, 1985)

The normalized left coprime factors $\tilde{N}$ *and* $\tilde{M}$ *of a transfer function* $G \in R$ *exist and are unique to within left multiplication by a unitary matrix.*

2.5.2 State-Space Properties of Normalized Coprime Factors

In later chapters it will be necessary to consider normalized coprime factors in a state-space framework. We now review a state-space construction of the normalized left coprime factors of a transfer function $G \in R$ with minimal state-space realization (A, B, C, D). To do this we have to introduce two Riccati equations associated with the minimal system (A, B, C, D):

The *Generalized Control Algebraic Riccati Equation* (GCARE) is given by,

$$(A - BS^{-1}D^*C)^*X + X(A - BS^{-1}D^*C) - XBS^{-1}B^*X + C^*R^{-1}C = 0,$$

and the *Generalized Filtering Algebraic Riccati Equation* (GFARE) is given by,

$$(A - BS^{-1}D^*C)Z + Z(A - BS^{-1}D^*C)^* - ZC^*R^{-1}CZ + BS^{-1}B^* = 0,$$

with $R = I + DD^*$ and $S = I + D^*D$.

(Appendix A gives a summary of the properties of GCARE and GFARE.)

It is well known that minimality of (A, B, C, D) is sufficient to ensure that symmetric and positive definite solutions to the GCARE and the GFARE exist, are unique and are the stabilizing ones. This is summarized in the following proposition:

Proposition 2.21

If (A, B, C, D) is controllable (observable), then there exists a unique solution $X = X^* > 0$, $(Z = Z^* > 0)$, to the GCARE (GFARE). If the ***control gain*** *F and the* ***filter gain*** *H are defined to be*

$$F \stackrel{\Delta}{=} -S^{-1}(D^*C + B^*X)$$

$$H \stackrel{\Delta}{=} -(BD^* + ZC^*)R^{-1},$$

then the eigenvalues of

$$(A + BF), \qquad (A + HC)$$

corresponding to these solutions, have strictly negative real parts.

The use of Riccati equations in this coprime factor context has its basis in the results of Nett *et al.*, (1984), who showed that left and right coprime factors of a nominal plant can be generated from a state-feedback/observer configuration.

The following proposition shows that the unique, positive definite solution to GFARE yields the filter gain required to form normalized coprime factors of a transfer function G. This result was first stated by Meyer and Franklin, (1987), for the case $D = 0$, and extended for $D \neq 0$ in Vidyasagar, (1988).

Proposition 2.22 (Vidyasagar, 1988)

Let (A, B, C, D) be a minimal state-space realization with associated transfer function $G(s) = C(sI - A)^{-1}B + D$. Further, if $\tilde{N}$ and $\tilde{M}$ have a state-space realization given by

$$[\tilde{N}, \tilde{M}] \stackrel{s}{=} \left[\begin{array}{c|cc} A + HC & B + HD & H \\ \hline R^{-1/2}C & R^{-1/2}D & R^{-1/2} \end{array}\right] \tag{2.9}$$

where

$$H \triangleq -(ZC^* + BD^*)R^{-1}$$

with Z the unique, positive definite solution to GFARE, and $R = I+DD^*$, then $(\tilde{N}, \tilde{M})$ is a normalized left coprime factorization of G.

The next proposition establishes the minimality of the realization in (2.9):

Proposition 2.23 (Ober and McFarlane, 1988; Meyer, 1988)

The realization in (2.9) is minimal if and only if the realization (A, B, C, D) is minimal.

This concludes the preliminary section on coprime factorizations. The reader is referred to Vidyasagar, (1985), for a comprehensive coverage of the subject.

2.6 Stability Results

This monograph is concerned with robust stability of closed-loop systems, and a necessary preliminary, therefore, is to clearly define what is meant by closed-loop stability. In Section 2.6.1, we introduce the concept of 'internal stability', and in the remainder of this work by 'closed-loop stable' we always mean internally stable. Section 2.6.2 then states two well known closed-loop stability results. Familiarity with basic open-loop and closed-loop stability requirements will be assumed.

2.6.1 Well-Posedness and Internal Stability

It is common in control system analysis to be concerned with *input-output stability.* A system is input-output stable if the output response to a bounded input signal is also bounded. However, a stronger notion of stability, is that of *internal stability*, discussed here, which precludes the possibility of any unbounded signals existing in a closed-loop system. In this section we will define internal stability and establish necessary and sufficient conditions for internal stability to be achieved. The results given here are standard, and can be found in Doyle and Chu, (1986).

Noting that many feedback systems can be expressed in a lower linear fractional transformation (LLFT) framework (see Figure 2.1, p13) we will give internal stability conditions in terms of this general feedback arrangement. For the purposes of this analysis we introduce external signals $\boldsymbol{v}_1$ and $\boldsymbol{v}_2$ as in Figure 2.3.

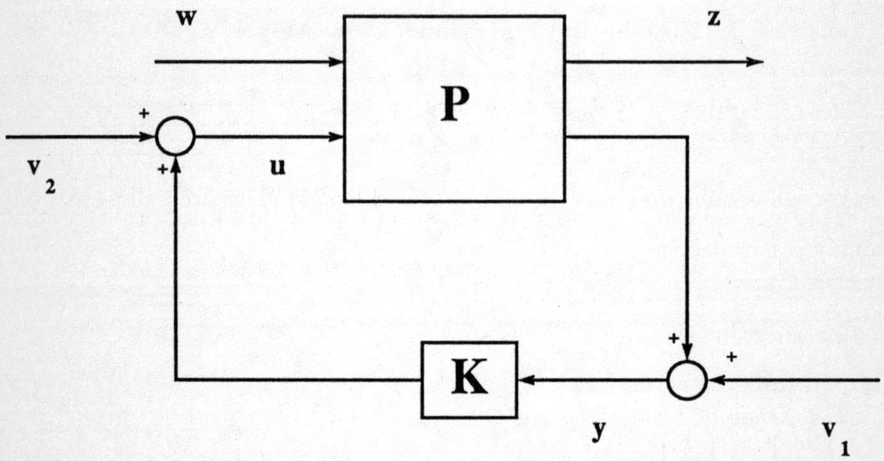

Figure 2.3 Generalized Feedback Arrangement

A useful preliminary concept is that of well posedness:

Definition 2.24

The system in Figure 2.1 is ***well-posed*** *if the transfer function matrix from* $\begin{Bmatrix} w \\ v_1 \\ v_2 \end{Bmatrix}$ *to* $\begin{Bmatrix} z \\ u \\ y \end{Bmatrix}$ *in Figure 2.3 exists and is proper.*

Assuming that $\boldsymbol{P}$ and $\boldsymbol{K}$ are proper transfer functions, a necessary and sufficient condition for well posedness is the following:

Lemma 2.25 (Doyle and Chu, 1986)

The system in Figure 2.1 is well-posed if and only if

$$(I - P_{22}K)(\infty)$$

is invertible, or equivalently, if and only if

$$(\tilde{M}V - \tilde{N}U)(\infty)$$

is invertible, where $(\tilde{N}, \tilde{M})$ *is a LCF of* P_{22} *and* (U, V) *is a RCF of* K.

We will assume from now on that all closed-loop systems are well-posed. Internal stability is now defined.

Definition 2.26

The system in Figure 2.1 is ***internally stable*** *if and only if the transfer function from* $\begin{Bmatrix} w \\ v_1 \\ v_2 \end{Bmatrix}$ *to* $\begin{Bmatrix} z \\ u \\ y \end{Bmatrix}$ *in Figure 2.3 is asymptotically stable.*

It is obvious that under some conditions (e.g. P_{11} unstable, $P_{12} = P_{21} = P_{22} = 0$) that *no* controller can achieve internal stability in Figure 2.1. To exclude such conditions we make the following assumption about the structure of P.

Assumption 2.27

Consider a minimal state-space representation of P *given by*

$$\left[\begin{array}{c:c} P_{11} & P_{12} \\ \hdashline P_{21} & P_{22} \end{array}\right] \stackrel{s}{=} \left[\begin{array}{c|c:c} A & B_1 & B_2 \\ \hline C_1 & D_{11} & D_{12} \\ \hdashline C_2 & D_{21} & D_{22} \end{array}\right],$$

then (A, B_2) *is assumed to be stabilizable and* (C_2, A) *is assumed to be detectable.*

This assumption says that if a controller, K, stabilizes P_{22} then it will also stabilize P. (That is, we have that $\mathcal{F}_L(P, K) \in RH_\infty$.) We will assume that Assumption 2.27 holds for all LLFT feedback arrangements. Under these conditions we can achieve a necessary and sufficient condition for the internal stability of the generalized feedback system in Figure 2.1.

Theorem 2.28 (Doyle and Chu, 1986)

The feedback system of Figure 2.1 is internally stable if and only if

(a) Assumption 2.27 holds

(b) $(\tilde{M}V - \tilde{N}U)^{-1}$ *is a unit in* RH_∞.

2.6.2 Closed-Loop Stability Results

We now state two results which are commonly used in closed-loop stability analysis. The Small Gain Theorem, first proposed by Zames (1966) is a conservative, but extremely simple, means for determining closed-loop stability. Consider the system in Figure 2.4.

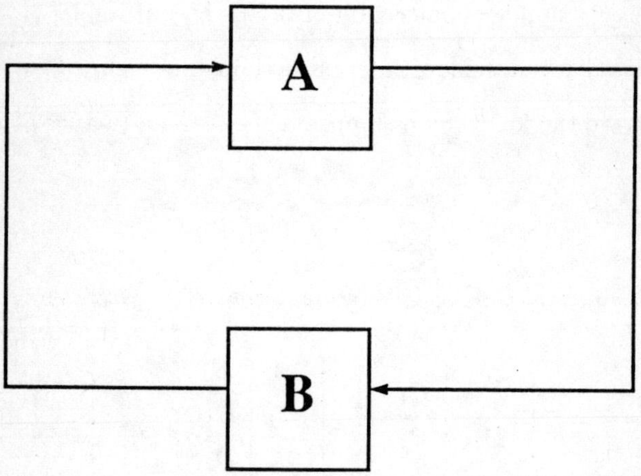

Figure 2.4 Closed-Loop System

Theorem 2.29 (Zames, 1966) ***'The Small Gain Theorem'***

The system of Figure 2.4, where $\boldsymbol{A}, \boldsymbol{B} \in RH_\infty$, *is internally stable if*

$$\overline{\sigma}(\boldsymbol{A}(j\omega)).\overline{\sigma}(\boldsymbol{B}(j\omega)) < 1 \textit{for all } \omega \in \Re.$$

Further, for $\boldsymbol{A}$ and $\boldsymbol{B}$ possibly unstable, we have the following result originally due to Rosenbrock, (1972), which extended the Nyquist stability criterion for scalar systems to the case of multivariable systems. The following version is from Vidyasagar, (1985, Lemma 7.4.8).

Theorem 2.30 ***'The Generalized Nyquist Criterion'***

If, in Figure 2.4, $\boldsymbol{A}$ *has* η_A *RHP poles, and* $\boldsymbol{B}$ *has* η_B *RHP poles, the closed-loop system is internally stable if and only if the Nyquist diagram of* $det(I - \boldsymbol{AB})(j\omega)$

(i) does not pass through the origin (i.e. $det(I - \boldsymbol{AB})(j\omega) \neq 0$ *for all* ω*.)*

(ii) encircles the origin $\eta_A + \eta_B$ *times anticlockwise.*

CHAPTER THREE

ROBUST STABILIZATION OF UNCERTAIN SYSTEMS

3.1 Introduction

The most important objective in control system design is stability. This is a minimum requirement for any design. Also, an unavoidable fact in design is that the assumed model of the system being considered for compensation (called the *nominal model*) will differ in behaviour from the actual system. This is because modelling physical systems for control design invariably involves a trade-off between simplicity of the model and its accuracy in matching the behaviour of the actual system. Whether the nominal model is obtained by theoretical or identification methods, it is often necessary to intentionally neglect complicated portions of a system so as to obtain a model which is suitable for mathematical manipulations. Differences between the actual plant and the nominal model are generally called *modelling errors* or *system uncertainty*, as there exists uncertainty as to how the model behaviour compares with the actual system behaviour.

Modelling errors can arise in many ways: For example, in design, it is generally necessary to consider a *linear* nominal model, while most physical systems are *non-linear* in some way. As a linear model cannot truly represent non linear behaviour, a modelling error is necessarily introduced. Also, even when a physical system is predominantly linear in behaviour, it is common to only include the most dominant modes in the nominal model, and again this introduces a modelling error. A further source of modelling error arises from inaccurate determination of model parameters, which is particularly common when the nominal model is determined using 'system identification' methods where parameters are estimated from input/output information.

In practice then, we design a controller for a nominal model whose behaviour is representative of the behaviour of the actual system. It is generally necessary therefore, to design a controller that not only stabilizes the nominal model, but which must also stabilize a *family* of systems which exist in an *uncertainty region* around the nominal model, where this 'family' is assumed to contain the actual system. The concept of two systems being in some way close to each other is commonly quantified in terms of the 'size' of the difference, or error, between their transfer functions. In this work we will use the L_∞ norm as a measure of modelling error size. An uncertainty region contains systems whose transfer functions have small errors when compared to the nominal system transfer function, and an uncertain plant model is described in terms of a nominal plant model and a specified uncertainty region.

The necessity to stabilize plant models which are uncertain motivates the idea of ***robust stability***: the ability of a closed-loop system to remain stable in the presence of modelling errors. Designing a controller which achieves robust stability in a closed-loop is a problem that has received a lot of attention in recent years, and in particular, the H_∞ control design approach has provided some promising results in the area of robust stabilization of plants with unstructured uncertainties. By unstructured uncertainty we mean uncertainty about which there is no information available except possibly a bound on its 'size' as a function of frequency.

In optimal H_∞ theory it is necessary to model plant uncertainty as a separate transfer function from the nominal plant model. Common approaches are to model uncertainty in a *multiplicative* or *additive* way with respect to the nominal plant (see Doyle and Stein, 1981 and Chen and Desoer, 1982) or as stable additive perturbations on the factors in a coprime factorization of the nominal plant (see Vidyasagar, 1984, 1985). Necessary and sufficient conditions for a controller to robustly stabilize a nominal plant with these forms of uncertainty can be established. (See Section 3.3.) This condition for robust stability involves a test on the H_∞ norm of a particular closed-loop transfer function, and hence the existence of a robustly stabilizing controller can be determined via the H_∞ optimization techniques originated by Zames, (1981). The above robust stability condition was observed by Kimura, (1984) in the SISO case, with MIMO extensions in Vidyasagar and Kimura, (1986), Glover, (1986a), and Verma *et al* (1986).

This chapter reviews the related subjects of system uncertainty and robust stability and introduces H_∞ methods in this context. In Section 3.2 it is shown that uncertainty can be modelled so that it both (a) represents the differences between a nominal system model and an uncertain system model, and (b) can be incorporated into closed-loop stability analysis. In Section 3.3 conditions for robust stability are derived. Finally in Section 3.4 it is shown that H_∞ optimization can be used to produce controllers which guarantee a degree of robust stability and in addition satisfy nominal performance objectives.

3.2 Modelling of Uncertainty

3.2.1 Background

For the presence of modelling error to play a part in controller analysis and design, it is necessary to construct some form of mathematical model to represent the error. The modelling should be based on the information available about the uncertainty, or, if this is unavailable, on reasonable assumptions about its magnitude.

A key assumption in all modelling of system errors is that the *uncertain* part of a process can be modelled *separately* from the *known* part of the process, as represented by the nominal model.

In current literature, modelling of model uncertainty is considered from two different viewpoints. First, in the *frequency domain*, the error or *perturbation* is considered to be a transfer function, separate from the system model. This approach is used as the basis for many analysis and design techniques. Doyle and Stein (1981) or Vidyasagar (1985) give good introductions to this approach. The second approach to modelling error is via the *state-space* representation of a process model, $G \stackrel{s}{=} (A, B, C, D)$, where variations in the A, B, C or D matrices are considered. This form of modelling is considered for example in the analyses of Hinrichsen and Pritchard (1986), Yedavalli (1985), Tahk and Speyer, (1987) and by Petersen, (1987b), and Khargonekar, Petersen and Zhou, (1987) in the 'quadratic stabilization' problem.

In this work we will restrict our attention to frequency domain uncertainty representations only.

3.2.2 Frequency Domain Error Modelling

In frequency-domain error modelling, uncertainty is assumed to be frequency dependent and complex. It is useful at this stage to make the distinction between 'structured' and 'unstructured' forms of uncertainty: *unstructured uncertainty* is uncertainty about which no information is available about its effects on a process, except that an upper bound on its 'size' or magnitude as a function of frequency can be estimated. *Structured uncertainty*, therefore, is uncertainty about which 'structural' information is available, which will typically restrict uncertainty to a section of a process model. It is desirable to 'structure' uncertainty if possible, as it restricts the class of uncertainty that the controller must be designed to combat. Unfortunately, it is uncommon to have structural information available, and it will be assumed in the remainder of this monograph that all uncertainty considered is unstructured.

In this section we will examine several unstructured uncertainty models which are now defined. Let G, G_Δ, be the transfer functions of the nominal and perturbed plant models respectively:

(i) A perturbation Δ_A is an *additive uncertainty* if

$$G_\Delta = G + \Delta_A. \tag{3.1}$$

(ii) A perturbation Δ_P is a *multiplicative (or proportional) uncertainty* if

$$G_\Delta = (I + \Delta_P)G. \tag{3.2}$$

Further, let $(\tilde{N}, \tilde{M})$ represent a left coprime factorization of G as defined in Definition 2.16, then

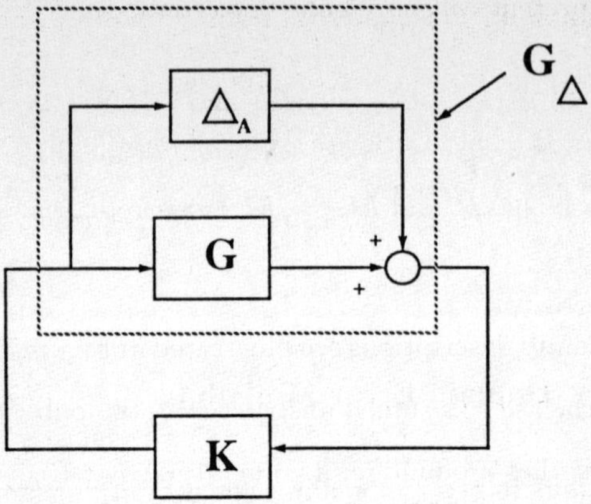

a) Additive Uncertainty

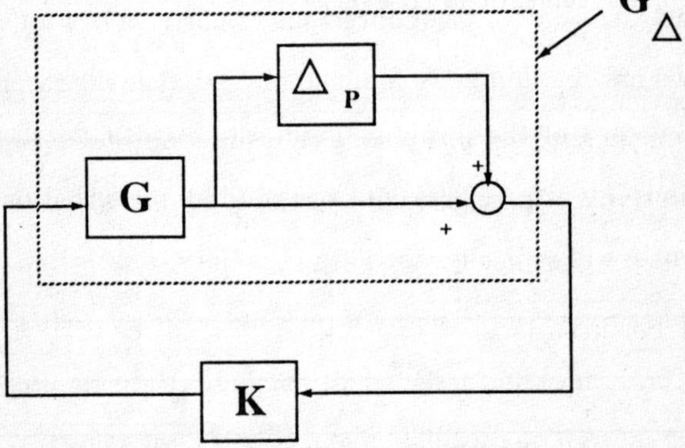

b) Multiplicative (Output) Uncertainty

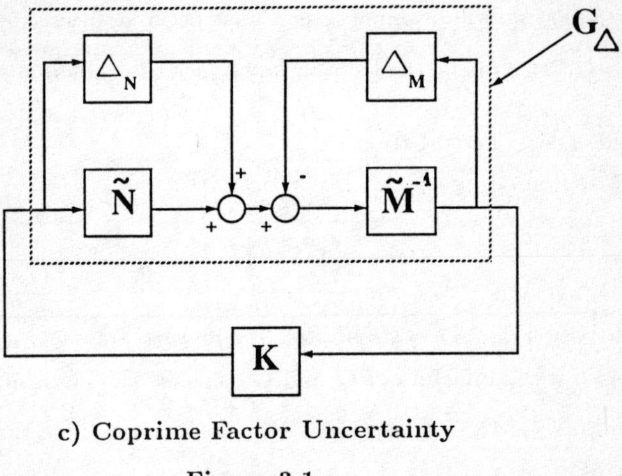

c) Coprime Factor Uncertainty

Figure 3.1

(iii) $[\Delta_N, \Delta_M]$ represent *coprime factor uncertainty* if

$$G_\Delta = (\tilde{M} + \Delta_M)^{-1}(\tilde{N} + \Delta_N). \tag{3.3}$$

where $\tilde{N}_\Delta = \tilde{N} + \Delta_N$, and $\tilde{M}_\Delta = \tilde{M} + \Delta_M$ represent a left coprime factorization of G_Δ.

These uncertainty descriptions are illustrated in Figure 3.1 (a)-(c). Note that as the uncertainty in each case is assumed unstructured, the only information available about the perturbation is that a bound on its magnitude over frequency can be estimated. (If Δ is a multivariable transfer function then an appropriate magnitude measure is the *maximum singular value*, denoted $\overline{\sigma}(\Delta)$.)

These classes are similar to the extent that it is always possible to represent a system error in *any* of the forms in Figure 3.1. Consider the SISO example where $G = \frac{1}{(s+1)}$ and $G_\Delta = \frac{(s-2)}{(s+1)(s-3)}$. Corresponding perturbations for each category in Figure 3.1 are:

(i) Additive Perturbations:

$$\Delta_A = G_\Delta - G = \frac{1}{(s+1)(s-3)}$$

(ii) Multiplicative (Proportional) Perturbations:

$$\Delta_P = G_\Delta G^{-1} - I = \frac{1}{(s-3)}$$

(iii) Coprime Factor Perturbations:

$$\begin{aligned} \Delta_N = \tilde{N}_\Delta - \tilde{N} &= \frac{(s-2)}{(s+3)(s+5)} - \frac{1}{(s+4)} = -\frac{(6s+23)}{(s+3)(s+4)(s+5)} \\ \Delta_M = \tilde{M}_\Delta - \tilde{M} &= \frac{(s+1)(s-3)}{(s+3)(s+5)} - \frac{(s+1)}{(s+4)} = -\frac{(7s+27)(s+1)}{(s+3)(s+4)(s+5)} \end{aligned}$$

In (iii) above, the factors of G and G_Δ can be chosen arbitrarily provided that $(\tilde{N}, \tilde{M})$ and $(\tilde{N}_\Delta, \tilde{M}_\Delta)$ are coprime. It is important to note that the perturbations Δ_N, Δ_M will always be stable since $\tilde{N}$, $\tilde{M}$, $\tilde{N}_\Delta$, $\tilde{M}_\Delta$ are stable by the definition of coprime factors. (This will be a benefit in the robust stability analysis in Section 3.3.)

Further confirmation of the equivalence of these uncertainty classes in representing uncertainty is that, for each case, the perturbed system model, G_Δ, can be written in the form of an ULFT (Upper Linear Fractional Transformation) as indicated in Figure 3.2, i.e.

$$G_\Delta = \mathcal{F}_U(P, \Delta),$$

where P is called the *Standard Plant* containing the nominal plant, G, and connections, and Δ represents the appropriate uncertainty model.

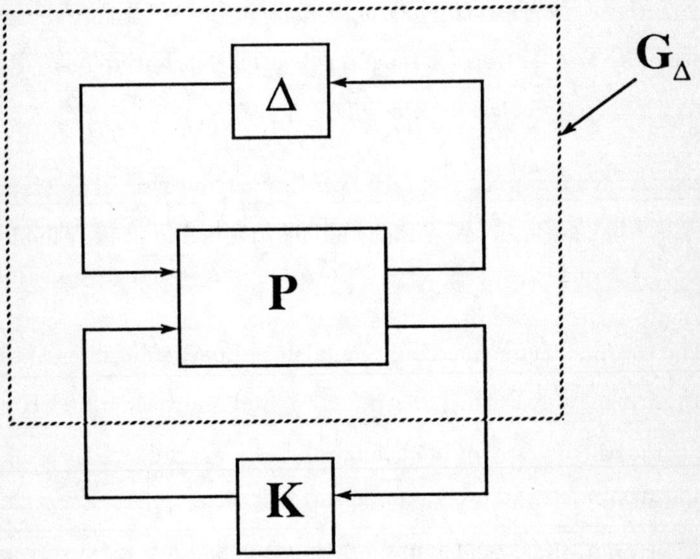

Figure 3.2 Generalized Uncertainty Model (with Feedback Controller)

The standard plants associated with the three uncertainty descriptions introduced earlier are

(i) Additive uncertainty

$$P = \left[\begin{array}{c:c} P_{11} & P_{12} \\ \hdashline P_{21} & P_{22} \end{array}\right] = \left[\begin{array}{c:c} 0 & I \\ \hdashline I & G \end{array}\right]; \quad (\Delta = \Delta_A). \tag{3.4}$$

(ii) Multiplicative uncertainty

$$P = \left[\begin{array}{c:c} P_{11} & P_{12} \\ \hdashline P_{21} & P_{22} \end{array}\right] = \left[\begin{array}{c:c} 0 & G \\ \hdashline I & G \end{array}\right]; \quad (\Delta = \Delta_P). \tag{3.5}$$

(iii) Coprime Factor Uncertainty

$$\boldsymbol{P} = \left[\begin{array}{c|c} \boldsymbol{P}_{11} & \boldsymbol{P}_{12} \\ \hline \boldsymbol{P}_{21} & \boldsymbol{P}_{22} \end{array}\right] = \left[\begin{array}{c|c} 0 & I \\ \tilde{M}^{-1} & G \\ \hline \tilde{M}^{-1} & G \end{array}\right]; \quad (\Delta = [\Delta_N, -\Delta_M]). \tag{3.6}$$

Now, using the multiplicative uncertainty case as an example, from the definition of an ULFT (see Definition 2.13), we can verify that this map represents a perturbed plant:

$$\begin{aligned} G_\Delta &= \mathcal{F}_U(\boldsymbol{P}, \Delta) \\ &= \boldsymbol{P}_{22} + \boldsymbol{P}_{21}\Delta(I - \boldsymbol{P}_{11}\Delta)^{-1}\boldsymbol{P}_{12} \\ &= G + I(I-0)^{-1}\Delta_P G \text{ by (3.5)} \\ &= G + \Delta_P G \quad \text{as required.} \end{aligned}$$

Also, note that for each uncertainty model, the nominal plant model is given by $G = \mathcal{F}_U(\boldsymbol{P}, 0)$.

The coprime factor uncertainty model is of particular interest in this work. It has been argued in Vidyasagar (1984), (1985), and Vidyasagar and Kimura, (1986), that this is an attractive way of modelling process uncertainty. It will be shown in Section 3.3 that, in robust stability analysis, coprime factor uncertainty allows a very general class of errors to be considered; more general than the corresponding classes allowed by additive and multiplicative models. In the next section we will consider the robust stabilization problem of finding a feedback controller which stabilizes $G_\Delta = \mathcal{F}_U(\boldsymbol{P}, \Delta)$ for all Δ in a particular class which must include $\Delta = 0$.

3.3 Robust Stability Analysis

3.3.1 Introduction

Robust stability is the ability of a controlled process to remain stable when uncertainty is present. In terms of the error models of Figure 3.1, a controller robustly stabilizes a process if it stabilizes any perturbed model G_Δ which is a combination of the nominal model, G and uncertainty model, $\Delta \in \mathcal{D}$, where $\mathcal{D}$ is a class of possible uncertainties which includes $\Delta = 0$. Hence as stated earlier, G_Δ really represents a family of models located around the nominal model.

In this section we establish sufficient conditions for a feedback controller $\boldsymbol{K}$ to robustly stabilize a plant $\boldsymbol{G}$. To do this we find the largest (in an L_∞ sense) perturbation Δ such that closed-loop stability is maintained. The L_∞ norm of this largest allowable perturbation is often called a *maximum stability margin.* We have chosen the L_∞ norm here, as it represents the maximum magnitude, over frequency, of the error Δ.

3.3.2 Frequency Domain Uncertainty Analysis

Frequency domain uncertainty analysis has received much attention in the 1980s largely due to the development of robust design techniques such as H_∞ Synthesis and Loop Transfer Recovery in LQG design (see Doyle and Stein 1979, Kwakernaak and Sivan, 1972). In this section we will derive explicit conditions for the robust stability of a closed-loop system.

In Section 3.2 we showed that a nominal plant, G, subject to additive, multiplicative or coprime factor uncertainty can be written as an ULFT:

$$G_\Delta = \mathcal{F}_U(P, \Delta) \tag{3.7}$$

and further, the nominal plant can be written as

$$G = \mathcal{F}_U(P, 0)$$

where P is a Standard Plant. We are going to use the closed-loop stability results in Theorems 2.29, 2.30 in combination with the perturbed system model in (3.7) to give sufficient conditions for robust stability. Before this, we define a class of 'permissible' perturbations :

Definition 3.1

A permissible perturbation, Δ, is one such that $\Delta \in \mathcal{D}_\epsilon$ where

$$\mathcal{D}_\epsilon \triangleq \mathcal{D}_{\mathcal{S}_\epsilon} \cup \mathcal{D}_{\mathcal{U}_\epsilon}$$

and

$$\mathcal{D}_{\mathcal{S}_\epsilon} \triangleq \{\Delta : \Delta \in RH_\infty; \quad \|\Delta\|_\infty < \epsilon\}$$

$$\mathcal{D}_{\mathcal{U}_\epsilon} \triangleq \{\Delta : \Delta \in RL_\infty; \quad \eta(\mathcal{F}_U(\boldsymbol{P},0)) = \eta(\mathcal{F}_U(\boldsymbol{P},\Delta)); \quad \|\Delta\|_\infty < \epsilon\}$$

where P is the standard plant, and $\eta(\bullet)$ denotes the number of closed RHP poles of a transfer function.

Remark 3.2

The set $\mathcal{D}_{\mathcal{S}_\epsilon}$ describes a set of *stable* bounded perturbations, and $\mathcal{D}_{\mathcal{U}_\epsilon}$ describes a set of bounded perturbations such that $\boldsymbol{G}$ and $\boldsymbol{G}_\Delta$ have the same number of closed RHP poles. We are therefore somewhat artificially restricting permissible perturbations to one of these two categories. The reason for this is that in the robust stability analysis to follow we can only treat these classes of perturbations. However, for the case of coprime factor uncertainty, we noted earlier that all possible perturbations are stable. Hence all coprime factor perturbations are contained in $\mathcal{D}_{\mathcal{S}_\epsilon}$ anyway, and the class of permissible perturbed systems is not restricted by Definition 3.1 in this case.

We now give sufficient conditions for robust stability for the uncertainty model in (3.7), which is a generalization of a result due to Vidyasagar (1985, Theorem 7.8).

Theorem 3.3

$\boldsymbol{K}$ stabilizes $\mathcal{F}_U(\boldsymbol{P},\Delta)$ in Figure 3.2 for all $\Delta \in \mathcal{D}_\epsilon$, and any standard plant $\boldsymbol{P}$ satisfying Assumption 2.27 if and only if

(a) $\boldsymbol{K}$ stabilizes $\mathcal{F}_U(\boldsymbol{P},0)$,

(b) $\|\mathcal{F}_L(P, K)\|_\infty \leq \epsilon^{-1}$.

Proof:

'if' - This part of the proof is split into two sections. We will consider the closed-loop stability of Figure 3.2 for all $\Delta \in \mathcal{D}_{\mathcal{S}_\epsilon}$ first, and then closed-loop stability for all $\Delta \in \mathcal{D}_{\mathcal{U}_\epsilon}$. (1) For the case $\Delta \in \mathcal{D}_{\mathcal{S}_\epsilon}$, we note that all perturbations are stable. Now by (a) we have that K stabilizes $\mathcal{F}_U(P, 0)$ and by Assumption 2.27, this guarantees internal stability when $\Delta = 0$ (i.e that $\mathcal{F}_L(P, K) \in RH_\infty$). Next noting by definition that $\|\Delta\|_\infty < \epsilon$, and by part (b) that $\|\mathcal{F}_L(P, K)\|_\infty \leq \epsilon^{-1}$, we have that

$$\|\mathcal{F}_L(P, K)\|_\infty . \|\Delta\|_\infty < 1.$$

By the Small Gain Theorem this is sufficient to guarantee stability. (Note that $\overline{\sigma}(\bullet)(j\omega) \leq \| \bullet \|_\infty$ for all $\omega \in \Re$.)

(2) Now consider the case $\Delta \in \mathcal{D}_{\mathcal{U}_\epsilon}$. As the condition $\Delta \in RH_\infty$ is not guaranteed, we cannot use the Small Gain Theorem as above. To prove this case then, note from the Generalized Nyquist Criterion (Theorem 2.30) that K internally stabilizes $\mathcal{F}_U(P, \Delta)$ if and only if

(i) $det(I - \mathcal{F}_U(P, \Delta)K)(j\omega) \neq 0$ for all ω.

(ii) the Nyquist diagram of $det(I - \mathcal{F}_U(P, \Delta)K)(j\omega)$ encircles the origin $\eta(\mathcal{F}_U(P, \Delta)) + \eta(K)$ times anticlockwise.

Next note that

$$\begin{aligned}
&det(I - \mathcal{F}_U(P, \Delta)K) \\
&= det(I - [P_{22} + P_{21}\Delta(I - P_{11}\Delta)^{-1}P_{12}]K) \\
&= det(I - P_{22}K) \\
&\qquad det\left((I - [P_{22} + P_{21}\Delta(I - P_{11}\Delta)^{-1}P_{12}]K)(I - P_{22}K)^{-1}\right) && (3.8) \\
&= det(I - P_{22}K)det(I - \Delta P_{11})^{-1} \\
&\qquad det(I - \Delta[P_{11} + P_{12}K(I - P_{22}K)^{-1}P_{21}]) && (3.9) \\
&= det(I - P_{22}K)det(I - \Delta P_{11})^{-1}det(I - \Delta\mathcal{F}_L(P, K)) && (3.10)
\end{aligned}$$

where (3.8) → (3.9) exploits the Schur formula for matrix determinants. (See Gantmacher, 1959, p46)

Further, noting that as $\boldsymbol{P}$ is stabilizable and detectable, we have that $\boldsymbol{P}$ has a RCF given by

$$\begin{bmatrix} \boldsymbol{P}_{11} & \boldsymbol{P}_{12} \\ \boldsymbol{P}_{21} & \boldsymbol{P}_{22} \end{bmatrix} = \begin{bmatrix} \boldsymbol{N}_1 \\ \boldsymbol{N}_2 \end{bmatrix} \boldsymbol{M}^{-1}$$

and therefore $\boldsymbol{P}_{11} = \boldsymbol{N}_1 \boldsymbol{M}^{-1} \begin{bmatrix} I \\ 0 \end{bmatrix}$ and $\boldsymbol{P}_{22} = \boldsymbol{N}_2 \boldsymbol{M}^{-1} \begin{bmatrix} 0 \\ I \end{bmatrix}$. Hence, using the Schur formula again we have

$$\begin{aligned} det(I - \Delta \boldsymbol{P}_{11})^{-1} &= det(I - \Delta \boldsymbol{N}_1 \boldsymbol{M}^{-1} \begin{bmatrix} I \\ 0 \end{bmatrix})^{-1} \\ &= det(\boldsymbol{M}) det(\boldsymbol{M} - \begin{bmatrix} I \\ 0 \end{bmatrix} \Delta \boldsymbol{N}_1)^{-1} \end{aligned}$$

and

$$det(I - \boldsymbol{P}_{22}\boldsymbol{K}) = det(\boldsymbol{M}^{-1}) det(\boldsymbol{M} - \begin{bmatrix} 0 \\ I \end{bmatrix} \boldsymbol{K}\boldsymbol{N}_2).$$

In (3.10) we then have

$$\begin{aligned} & det(I - \mathcal{F}_U(\boldsymbol{P}, \Delta)\boldsymbol{K}) \\ &= det(\boldsymbol{M} - \begin{bmatrix} I \\ 0 \end{bmatrix} \Delta \boldsymbol{N}_1)^{-1} det(\boldsymbol{M} - \begin{bmatrix} 0 \\ I \end{bmatrix} \boldsymbol{K}\boldsymbol{N}_2) det(I - \Delta \mathcal{F}_L(\boldsymbol{P}, \boldsymbol{K})) \\ &\neq 0 \end{aligned}$$

as $det(\boldsymbol{M} - \begin{bmatrix} 0 \\ I \end{bmatrix} \boldsymbol{K}\boldsymbol{N}_2) \neq 0$ by internal stability (condition (a)), and by Assumption 2.27 (which guarantees that $(\boldsymbol{N}_2, \boldsymbol{M})$ is RCF of $\boldsymbol{P}_{22}$); $det(\boldsymbol{M} - \begin{bmatrix} I \\ 0 \end{bmatrix} \Delta \boldsymbol{N}_1)^{-1} \neq 0$ because $\boldsymbol{M}$, $\boldsymbol{N}_1$ are asymptotically stable, and $\|\Delta\|_\infty$ is bounded; and finally, $det(I - \Delta \mathcal{F}_L(\boldsymbol{P}, \boldsymbol{K})) \neq 0$ by condition (b).

Hence we have shown that condition (i) of the Generalized Nyquist Criterion is satisfied for all $\Delta \in \mathcal{D}_{\mathcal{U}_\epsilon}$. To show that condition (ii) is also satisfied, note that as $det(I - \mathcal{F}_U(\boldsymbol{P}, \Delta)\boldsymbol{K})(j\omega) \neq 0$ for all $\Delta \in \mathcal{D}_{\mathcal{U}_\epsilon}$ (including $\Delta = 0$) then the Nyquist diagram of $det(I - \mathcal{F}_U(\boldsymbol{P}, \Delta)\boldsymbol{K})$ encircles the origin $\eta(\mathcal{F}_U(\boldsymbol{P}, 0)) + \eta(\boldsymbol{K})$ times anticlockwise for all $\Delta \in \mathcal{D}_{\mathcal{U}_\epsilon}$. Further as we assume $\eta(\mathcal{F}_U(\boldsymbol{P}, \Delta)) = \eta(\mathcal{F}_U(\boldsymbol{P}, 0))$, condition (ii) is now satisfied, and this section of the proof of Theorem 3.3 is complete.

'only if' - Firstly note that it is obvious that condition a) must be a necessary condition for internal stability of the perturbed system. To show that b) is also a necessary condition, we must show that $\|\mathcal{F}_L(P,K)\|_\infty > \epsilon^{-1}$ necessarily implies instability. This implies $\epsilon\overline{\sigma}(\mathcal{F}_L(P,K))(j\omega_o)) > 1$ for some ω_o, and necessity is proven if, given this condition, there is a $\Delta_u \in \mathcal{D}_\epsilon$ which destabilizes the closed-loop system. To show this, we must construct a $\Delta_u \in \mathcal{D}_\epsilon$ such that $det(I - \Delta_u\mathcal{F}_L(P,K))(j\omega_o) = 0$, as by (3.10) this implies $det(I - \mathcal{F}_U(P,\Delta_u)K)(j\omega_o) = 0$, which implies instability. Vidyasagar, (1985, pp 275-278), shows that such a Δ_u will exist given $\epsilon\overline{\sigma}(\mathcal{F}_L(P,K))(j\omega_o)) > 1$ for some ω_o, and outlines a suitable construction where Δ_u is stable, $\eta(\mathcal{F}_U(P,0)) = \eta(\mathcal{F}_U(P,\Delta_u))$, and $\|\Delta_u\|_\infty < \epsilon$, i.e. $\Delta_u \in (\mathcal{D}_{\mathcal{S}_\epsilon} \cap \mathcal{D}_{\mathcal{U}_\epsilon})$. The reader is referred to this text for details of this construction. This proves necessity and hence completes the proof of Theorem 3.3 ■

Remark 3.4

Note that Theorem 3.3 gives a very general condition for robust stability and can be applied to any perturbed system that can be written as an ULFT, and not just the three uncertainty models that we have considered in this chapter. We will use this result in section 3.4 to construct a standard framework for robust controller design.

Before moving on to controller synthesis problems, we will apply the robustness analysis conditions of Theorem 3.3 to the specific cases where the nominal plant is subject to additive, multiplicative and coprime factor uncertainty. These results have been previously stated in Vidyasagar, (1985) and Vidyasagar and Kimura, (1986). We first consider additive uncertainty.

Corollary 3.5

K stabilizes $G_\Delta = G + \Delta_A$ in Figure 3.1(a), for all $\Delta \in \mathcal{D}_\epsilon$ if and only if

(a) K stabilizes G.

(b) $\|K(I - GK)^{-1}\|_\infty \leq \epsilon^{-1}$.

Proof:

The proof is immediate from Theorem 3.3. G_Δ can be written

$$G_\Delta = G + \Delta_A = \mathcal{F}_u(P, \Delta_A)$$

where

$$P = \left[\begin{array}{c|c} P_{11} & P_{12} \\ \hline P_{21} & P_{22} \end{array}\right] = \left[\begin{array}{c|c} 0 & I \\ \hline I & G \end{array}\right].$$

The result follows by noting that

$$\mathcal{F}_L(P, K) = K(I - GK)^{-1}.$$

■

Next we state the analogous multiplicative uncertainty result. (The proof is omitted as it is similar to above.)

Corollary 3.6

K stabilizes $G_\Delta = G + \Delta_P G$ in Figure 3.1(b), for all $\Delta_P \in \mathcal{D}_\epsilon$ if and only if

(a) K stabilizes G.

(b) $\|GK(I - GK)^{-1}\|_\infty \leq \epsilon^{-1}$.

We lastly give a robust stability result for coprime factor uncertainty. Note that in this case, we can restrict uncertainty to the class $\mathcal{D}_{S\epsilon}$ without loss of generality.

Corollary 3.7

K stabilizes $G_\Delta = (\tilde{M} + \Delta_M)^{-1}(\tilde{N} + \Delta_N)$ in Figure 3.1(c) for all $\Delta_C \triangleq [\Delta_N, \Delta_M] \in \mathcal{D}_{S\epsilon}$ if and only if

(a) K stabilizes G.

(b) $\left\| \begin{bmatrix} K(I - GK)^{-1}\tilde{M}^{-1} \\ (I - GK)^{-1}\tilde{M}^{-1} \end{bmatrix} \right\|_\infty \leq \epsilon^{-1}.$

Proof:

First note that

$$G_\Delta = \mathcal{F}_U(P, \Delta)$$

and recall that the standard plant for this uncertainty model is

$$P = \left[\begin{array}{c|c} 0 & I \\ \tilde{M}^{-1} & G \\ \hline \tilde{M}^{-1} & G \end{array}\right].$$

Now, as

$$\mathcal{F}_L(P, K) = \begin{bmatrix} K(I - GK)^{-1}\tilde{M}^{-1} \\ (I - GK)^{-1}\tilde{M}^{-1} \end{bmatrix}$$

the result follows from the proof of Theorem 3.3. ■

Remark 3.8

An immediate benefit of the coprime factor uncertainty approach is that the uncertainty class is not restricted to perturbations which preserve the number of RHP poles of the plant. This enables a much greater confidence in the robust stability conditions obtained, as a wider class of perturbations is being considered. A simple example is $G = \frac{1}{(s+\kappa)}$ and $G_\Delta = \frac{1}{(s-\kappa)}$. Obviously, for κ small, the two models are very similar, and in fact, both will be stabilized by a single constant feedback K, if $K < -\kappa$. However, for the additive and multiplicative uncertainty classes, (Δ_A and Δ_P respectively) this is not a permissible perturbation, as Δ_A and Δ_P are both unstable, and the number of closed RHP poles in the perturbed plant is different to that of the nominal plant. In contrast, for the coprime factor uncertainty case, it is simple to show that the perturbations Δ_N, and Δ_M are stable, and therefore $[\Delta_N, \Delta_M] \in \mathcal{D}_{S_\epsilon}$ for some ϵ. Hence for the coprime factor uncertainty class G_Δ is a permissible perturbed plant.

Remark 3.9

A further benefit of the coprime factor approach is that, as $\tilde{N}$ and $\tilde{M}$ are asymptotically stable functions, $\overline{\sigma}(\tilde{N})$ and $\overline{\sigma}(\tilde{M})$ are bounded, and we therefore have the following guarantees for allowable additive and multiplicative uncertainty: From Corollary

3.7, we have

$$\left\| \begin{bmatrix} K \\ I \end{bmatrix} (I - GK)^{-1}\tilde{M}^{-1} \right\|_\infty \leq \epsilon^{-1}$$

$$\Rightarrow \overline{\sigma}(K(I - GK)^{-1}\tilde{M}^{-1}) \leq \epsilon^{-1}$$

$$\Rightarrow \overline{\sigma}(K(I - GK)^{-1}) \leq \epsilon^{-1}\overline{\sigma}(\tilde{M})$$

$$\Rightarrow \|K(I - GK)^{-1}\|_\infty \leq \epsilon^{-1}\|\tilde{M}\|_\infty$$

and similarly

$$\|GK(I - GK)^{-1}\|_\infty \leq \epsilon^{-1}\|\tilde{M}\|_\infty + 1.$$

We will exploit this result in Chapter 6.

In the next section we will show that the results in Theorem 3.3 and Corollaries 3.5 - 3.7 can be applied to robust stabilization problems which can be solved using H_∞ optimization.

3.4 Robust Controller Design Using H_∞ Optimization

3.4.1 The Robust Stabilization Problem

The results in Section 3.3 will now be exploited to generate controllers which guarantee a level of robust stability in a closed-loop system. We call such controllers *robustly stabilizing controllers.* Using the general framework of Figure 3.2, we can use the results of Theorem 3.3 to state the following robust stabilization problem:

Proposition 3.10 ***'The Robust Stabilization Problem'***

The largest strictly positive number $\epsilon = \epsilon_{max}$, such that, for all $\Delta \in \mathcal{D}_\epsilon$, there exists a single controller which stabilizes $\mathcal{F}_U(P, \Delta)$ is given by

$$\epsilon_{max} = \left(\inf_K \|\mathcal{F}_L(P, K)\|_\infty \right)^{-1}$$

where K is chosen from all controllers which stabilize $\mathcal{F}_U(P, 0)$.

In Proposition 3.10 we have written the Robust Stabilization problem as an H_∞ *optimization problem* where a controller K is chosen to minimize the H_∞ norm of a closed-loop transfer function, under the constraint that it must also stabilize the nominal plant. The stabilizing controller achieving the inf is called the *optimal*, or *optimally robust*, controller.

Remark 3.11

Generally, the uncertainty size will be known to depend on frequency and this can be specified by introducing a *weighting function*, W, and by requiring that $\|\Delta W^{-1}\|_\infty < \epsilon$ in the uncertainty class $\mathcal{D}_\epsilon$ in Definition 3.1 (rather than $\|\Delta\|_\infty < \epsilon$), in which case robust stability is then implied by requiring $\|W\mathcal{F}_L(P,K)\|_\infty \leq \epsilon^{-1}$. $\mathcal{F}_L(P,K)$ can then be rewritten as $\mathcal{F}_L(P',K)$ where W is included in an appended Standard Plant, P', by the construction

$$P' = \left[\begin{array}{c|c} WP_{11} & WP_{12} \\ \hline P_{21} & P_{22} \end{array}\right].$$

We have now shown that a robust stabilization problem can be written as an H_∞ optimization problem, and in Sections 3.4.3 and 3.4.4 we will discuss techniques for solving H_∞ optimization problems. However, before we do this we will examine how to also incorporate performance specifications into an H_∞ framework. This is important in robust controller design, as in general, a satisfactory controller cannot be obtained by considering robust stability objectives alone.

3.4.2 Nominal Performance Problems

In the earlier sections of this chapter we have shown that the H_∞ norm of a closed-loop transfer function is a particularly useful measure to minimize because it enables robust stability guarantees to be made. A further property of the H_∞ norm was given in Theorem 2.12: The H_∞ norm of a transfer function matrix represents the maximum energy in the output signal from that transfer function due to any input of unit energy. Hence minimizing the H_∞ norm of a transfer function is equivalent to minimizing the energy in the output signal due to inputs of unit energy.

For closed-loop systems, this property can be readily applied to common performance problems such as disturbance rejection or tracking error minimization. By minimizing the H_∞ norm of a related closed-loop transfer function, a level of nominal performance can be guaranteed. For example, note that $G(I-GK)^{-1}$ is the transfer function from input disturbances to the plant output. Then, if a stabilizing controller, K, is chosen so that $\|G(I-GK)^{-1}\|_\infty$ is minimized, the energy of the plant output signal due to input disturbances (of bounded energy) will be minimized. Similarly, note that $(I-GK)^{-1}$ is the transfer function from output disturbances to plant output. Hence, minimizing $\|(I-GK)^{-1}\|_\infty$ (the H_∞ norm of the sensitivity function) minimizes the energy of the output signal due to disturbances (of bounded energy) on the plant output.

We could rewrite either of the above minimization problems in a standard framework as in Proposition 3.10, where we want to find

$$\inf_{K \text{ stabilizing}} \|\mathcal{F}_L(P,K)\|_\infty .$$

For example, the standard plant P associated with the sensitivity minimization problem would be

$$P = \left[\begin{array}{c:c} P_{11} & P_{12} \\ \hdashline P_{21} & P_{22} \end{array}\right] = \left[\begin{array}{c:c} I & G \\ \hdashline I & G \end{array}\right].$$

A table of common performance and robust stability objectives and their associated closed-loop transfer functions is given in Table 3.1.

Again, we can use frequency weighting on the appropriate closed-loop transfer functions to reflect the relative magnitude and importance of disturbances and errors.

FUNCTION	INTERPRETATION
$\overline{\sigma}((I-GK)^{-1})$	- gain from output disturbance to plant output - gain from reference signal to tracking error
$\overline{\sigma}((I-KG)^{-1})$	- gain from input disturbance to plant input
$\overline{\sigma}(K(I-GK)^{-1})$	- gain from output disturbance to controller input - $1/\overline{\sigma}(K(I-GK)^{-1})$ indicates the maximum allowable additive plant perturbation for closed-loop stability
$\overline{\sigma}((I-GK)^{-1}G)$	- gain from input disturbance to plant output - $1/\overline{\sigma}((I-GK)^{-1}G)$ indicates the maximum allowable additive controller perturbation for closed-loop stability
$\overline{\sigma}(GK(I-GK)^{-1})$	- gain from controller input disturbance to plant output - $1/\overline{\sigma}(GK(I-GK)^{-1})$ indicates the maximum allowable output multiplicative plant perturbation for closed-loop stability.
$\overline{\sigma}(K(I-GK)^{-1}G)$	- gain from input disturbance to controller output - $1/\overline{\sigma}(K(I-GK)^{-1}G)$ indicates the maximum allowable input multiplicative plant perturbation for closed-loop stability.

Common Closed-Loop Transfer Function Objectives

Table 3.1

3.4.3 H_∞ Problem Specification

We now have means of representing both robust stability and performance objectives in the same H_∞ minimization framework. An H_∞ optimization problem specification then typically combines a number of frequency weighted closed-loop transfer functions and minimizes the H_∞ norm of the composite transfer function matrix. The appropriate selection of weights over frequency and between objectives is not straightforward, and tends to be developed for each specific example.

It is possible to combine many such objectives into the form of a single linear fractional transformation, $\mathcal{F}_L(P, K)$, which is to be minimized by choice of K. (See Francis and Doyle, 1987, for some examples.) For completeness, the H_∞ optimization problem can be formally stated as:

Problem 3.12 ***'The H_∞ Optimization Problem'***

Find

$$\inf_{K} \|\mathcal{F}_L(P, K)\|_\infty = \gamma_{min}$$

where K is chosen over all controllers which internally stabilize $\mathcal{F}_U(P, 0)$.

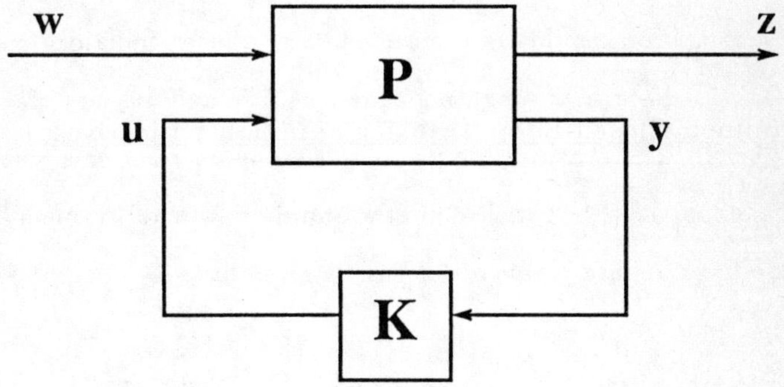

Figure 3.3 H_∞ Objective Function

Referring to Figure 3.3, $\mathcal{F}_L(P, K)$ represents the transfer function from an exogeneous input w to an output signal z, and P is the standard plant associated with a particular combination of objectives. (The standard plant P includes the nominal plant G, weighting functions, and interconnections to make the required closed-loop from $w \rightarrow z$ in Figure 3.3.) This combination of objectives into a single H_∞ problem is best illustrated by an example.

Example 3.13

An example of a common H_∞ design problem in that of minimizing the H_∞ norm of a combination of the closed-loop transfer functions $(I - GK)^{-1}$, and $K(I - GK)^{-1}$ by choosing an appropriate stabilizing controller, K. From Section 3.4.2 we note that $(I - GK)^{-1}$ is a performance related transfer function, while $K(I - GK)^{-1}$ was shown in Corollary 3.5 to give a robust stability guarantee. Note that performance considerations are typically most important at low frequencies while robust stability considerations are typically most important at high frequency. Following this argument, we then choose stable, minimum phase transfer functions W_2 (resp. W_1) to appropriately weight $(I - GK)^{-1}$ (resp. $K(I - GK)^{-1}$) and combining the weighted objectives, the H_∞ optimization problem is: Find

$$\inf_{K \text{ Stabilizing}} \left\| \begin{bmatrix} W_2(I - GK)^{-1} \\ W_1 K(I - GK)^{-1} \end{bmatrix} \right\|_\infty$$

where we are minimizing the H_∞ norm of the transfer function from w to $\left\{ \begin{smallmatrix} z_1 \\ z_2 \end{smallmatrix} \right\}$ in Figure 3.4(a). Note that weighting functions W_1 and W_2 are chosen so that the minimization of each transfer function is most important over a selected frequency range. This problem can be rewritten in the standard form of Problem 3.12 (see Figure 3.4b), where the standard plant in this case is given by

$$P = \left[\begin{array}{c|c} P_{11} & P_{12} \\ \hline P_{21} & P_{22} \end{array} \right] = \left[\begin{array}{c|c} W_2 & W_2 G \\ 0 & W_1 \\ \hline I & G \end{array} \right].$$

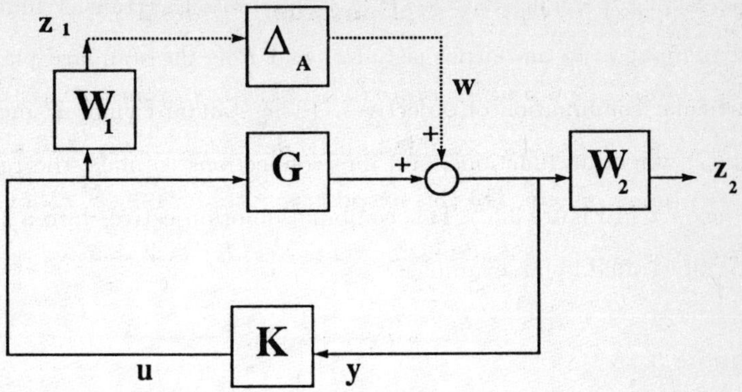

a) Regular Feedback Arrangement

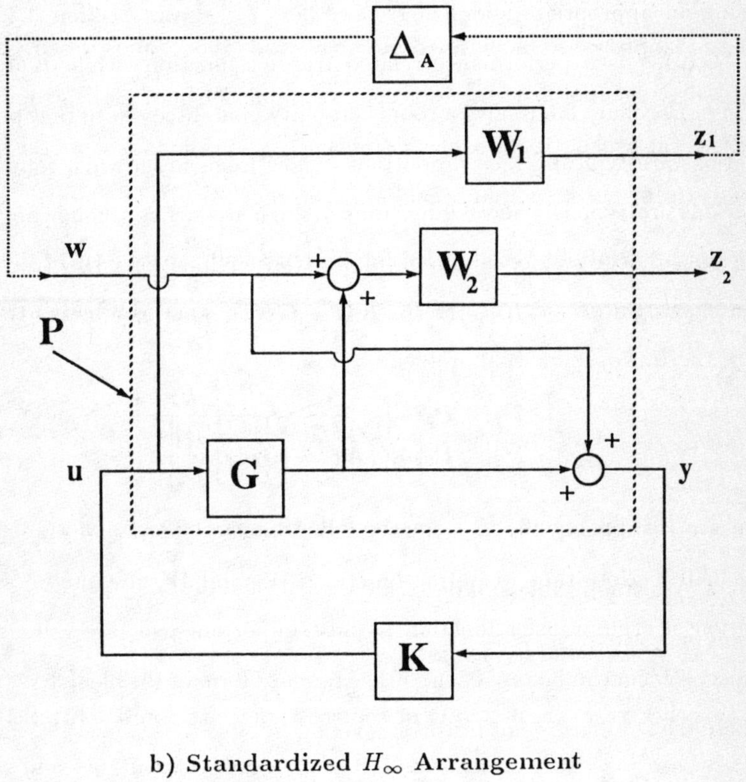

b) Standardized H_∞ Arrangement

Figure 3.4

3.4.4 Solution to the H_∞ Optimization Problem

We now present a solution procedure for the H_∞ optimization problem of Problem 3.12. In general, it is not possible to solve for γ_{min} *exactly*, and we will consider an iterative procedure here. For this purpose we introduce the so called *Sub-Optimal H_∞ Problem*, that of finding a stabilizing controller, K, such that for $\gamma > \gamma_{min}$,

$$\|\mathcal{F}_L(P, K)\|_\infty \leq \gamma. \tag{3.11}$$

Glover and Doyle, (1988) have stated necessary and sufficient conditions for the existence of a stabilizing controller solving (3.11), and parametrize all such controllers. We review this work next, and then outline an iterative procedure for estimating γ_{min}. The derivation of this procedure is outlined in Doyle *et al* (1988), and presents significant simplifications over earlier procedures (see, for example, Doyle, 1984, Foo and Postlethwaite, 1984, Kwakernaak, 1985, Safonov, Jonckheere, Verma and Limebeer, 1987, and Francis, 1987) in terms of computational requirements.

We begin this solution procedure by noting that a state-space form of the Standard Plant $P \in R$, with dimensions $(p_1 + p_2) \times (m_1 + m_2)$, with MacMillan degree n, is given by

$$P = \left[\begin{array}{c:c} P_{11} & P_{12} \\ \hdashline P_{21} & P_{22} \end{array}\right] \stackrel{s}{=} \left[\begin{array}{c|c:c} A & B_1 & B_2 \\ \hline C_1 & D_{11} & D_{12} \\ \hdashline C_2 & D_{21} & D_{22} \end{array}\right] \tag{3.12}$$

In order to use this procedure, we make the following assumptions on P:

(A1) (A, B_2, C_2) is stabilizable and detectable.

(A2) $rank(D_{12}) = m_2$, $rank(D_{21}) = p_2$

(A3) Referring to Figure 3.3, a scaling of controller output, u and controller input, y, together with a unitary transformation of exogeneous input, w, and output, z, enables us to assume without loss of generality that

$$D_{12} = \begin{bmatrix} 0 \\ I_{m_2} \end{bmatrix}; D_{21} = [0, I_{p_2}]; \text{ and } D_{11} = \begin{bmatrix} D_{1111} & D_{1112} \\ D_{1121} & D_{1122} \end{bmatrix}$$

is dimensioned compatibly with D_{12} and D_{21}.

(A4) $D_{22} = 0$.

(A5) $rank\begin{bmatrix} A - j\omega I & B_2 \\ C_1 & D_{12} \end{bmatrix} = n + m_2$ for all $\omega \in \Re$.

(A6) $rank\begin{bmatrix} A - j\omega I & B_1 \\ C_2 & D_{21} \end{bmatrix} = n + p_2$ for all $\omega \in \Re$.

Assumption (A1) guarantees the existence of a stabilizing controller, assumptions (A2), (A4), (A5) and (A6) guarantee that the solution to the problem is well posed, and (A3) is convenient for this solution procedure. We will show in Remark 3.15 that the $D_{22} \neq 0$ case can, in general, also be solved.

Now define

$$R \triangleq D_{1\bullet}^* D_{1\bullet} - \begin{bmatrix} \gamma^2 I_{m_1} & 0 \\ 0 & 0 \end{bmatrix} \text{ where } D_{1\bullet} \triangleq [D_{11}, D_{12}]$$

$$\tilde{R} \triangleq D_{\bullet 1} D_{\bullet 1}^* - \begin{bmatrix} \gamma^2 I_{p_1} & 0 \\ 0 & 0 \end{bmatrix} \text{ where } D_{\bullet 1} \triangleq \begin{bmatrix} D_{11} \\ D_{21} \end{bmatrix}$$

The stabilizing solution to an Algebraic Riccati Equation (ARE) is denoted via its associated Hamiltonian matrix as

$$X = Ric\left\{ \begin{matrix} A & -P \\ -Q & -A^* \end{matrix} \right\}, P = P^*, Q = Q^*,$$

where this implies $X = X^*$ and

$$XA + A^*X - XPX + Q = 0, \quad Re[\lambda_i(A - PX)] < 0.$$

Now, define X_∞, Y_∞ as ARE solutions (assuming solutions exist) defined by

$$X_\infty = Ric\left\{ \begin{bmatrix} A & 0 \\ -C_1^*C_1 & -A^* \end{bmatrix} - \begin{bmatrix} B \\ -C_1^* D_{1\bullet} \end{bmatrix} R^{-1} [D_{1\bullet}^* C_1, B^*] \right\} \tag{3.13}$$

$$Y_\infty = Ric\left\{ \begin{bmatrix} A^* & 0 \\ -B_1 B_1^* & -A \end{bmatrix} - \begin{bmatrix} C^* \\ -B_1 D_{\bullet 1}^* \end{bmatrix} \tilde{R}^{-1} [D_{\bullet 1} B_1^*, C] \right\} \tag{3.14}$$

and define 'state-feedback' and 'output-injection' matrices by

$$F_I = \begin{bmatrix} F_{11} \\ F_{12} \\ F_2 \end{bmatrix} = -R^{-1}[D_{1\bullet}^* C_1 + B^* X_\infty]$$

$$H_I = [\, H_{11} \quad H_{12} \quad H_2 \,]$$

$$= -[B_1 D_{\bullet 1}^* + Y_\infty C^*]\tilde{R}^{-1}.$$

where F_I, H_I are partitioned conformally with $D_{1\bullet}^*$ and $D_{\bullet 1}^*$ respectively. We will now give a state-space formula for all stabilizing controllers achieving $\|\mathcal{F}_L(\boldsymbol{P}, \boldsymbol{K})\|_\infty \leq \gamma$.

Theorem 3.14 (Glover and Doyle, 1988)

For the system given in (3.12) satisfying assumptions (A1) - (A6)

(a) There exists an internally stabilizing controller K such that $\|\mathcal{F}_L(P,K)\|_\infty < \gamma$ if and only if

(i) $\gamma > \gamma_l$; $\gamma_l = max(\overline{\sigma}[D_{1111}, D_{1112}], \overline{\sigma}[D^*_{1111}, D^*_{1121}])$ and

(ii) there exists $X_\infty \geq 0$, $Y_\infty \geq 0$ satisfying (3.13) and (3.14) respectively, such that $\rho(X_\infty Y_\infty) < \gamma^2$.

(b) Given that the conditions in (a) are satisfied, then all rational, internally stabilizing controllers K satisfying $\|\mathcal{F}_L(P,K)\|_\infty \leq \gamma$, for $\gamma > \gamma_{min}$, are given by $K = \mathcal{F}_L(K_a, \Phi)$ for arbitrary $\Phi \in RH_\infty^{m_2 \times p_2}$ such that $\|\Phi\|_\infty \leq \gamma$, where

$$K_a = \left[\begin{array}{c|c} K_{11} & K_{12} \\ \hline K_{21} & K_{22} \end{array}\right] \stackrel{s}{=} \left[\begin{array}{c|c:c} \hat{A} & \hat{B}_1 & \hat{B}_2 \\ \hline \hat{C}_1 & \hat{D}_{11} & \hat{D}_{12} \\ \hdashline \hat{C}_2 & \hat{D}_{21} & 0 \end{array}\right],$$

$$\hat{D}_{11} = -D_{1121}D^*_{1111}(\gamma^2 I - D_{1111}D^*_{1111})^{-1}D_{1112} - D_{1122}\ ,$$

$\hat{D}_{12} \in \Re^{m_2 \times m_2}$, $\hat{D}_{21} \in \Re^{p_2 \times p_2}$ are any matrices satisfying

$$\hat{D}_{12}\hat{D}^*_{12} = I - D_{1121}(\gamma^2 I - D_{1111}D^*_{1111})^{-1}D^*_{1121}$$
$$\hat{D}^*_{21}\hat{D}_{21} = I - D^*_{1112}(\gamma^2 I - D_{1111}D^*_{1111})^{-1}D_{1112}$$

and

$$\hat{B}_2 = (B_2 + H_{12})\hat{D}_{12}$$
$$\hat{C}_2 = -\hat{D}_{21}(C_2 + F_{12})Z$$
$$\hat{B}_1 = -H_2 + \hat{B}_2\hat{D}_{12}^{-1}\hat{D}_{11}$$
$$\hat{C}_1 = F_2 Z + \hat{D}_{11}\hat{D}_{21}^{-1}\hat{C}_2$$
$$\hat{A} = A + H_I C + \hat{B}_2\hat{D}_{12}^{-1}\hat{C}_1$$

where

$$Z = (I - \gamma^{-2}Y_\infty X_\infty)^{-1}.$$

Remark 3.15

If the assumption (A4) is removed, i.e. $D_{22} \neq 0$, the solution in Theorem 3.14 is followed setting $D_{22} = 0$, and then, assuming that $det(I + \hat{D}_{11} D_{22}) \neq 0$, all stabilizing controllers achieving $\|\mathcal{F}_L(P, K)\|_\infty \leq \gamma$ are given by

$$\tilde{K} = \mathcal{F}_L(\tilde{K}_a, \Phi), \text{ for } \Phi \in RH_\infty^{m_2 \times p_2}, \|\Phi\|_\infty \leq \gamma$$

where

$$\tilde{K}_a = \left[\begin{array}{c|c} \hat{A} - \hat{B}(I - M)\hat{D}^{-1}\hat{C} & \hat{B}M \\ \hline \tilde{M}C & \hat{D}M \end{array}\right]$$

and

$$M = \left(I + \begin{bmatrix} D_{22} & 0 \\ 0 & 0 \end{bmatrix} \hat{D}\right)^{-1}$$

$$\tilde{M} = \left(I + \hat{D}\begin{bmatrix} D_{22} & 0 \\ 0 & 0 \end{bmatrix}\right)^{-1}$$

where all terms are as defined in Theorem 3.14.

Remark 3.16

In order to obtain a controller that approximately achieves the minimum value of γ (to within a tolerance μ), the following iterative procedure can be used, employing conditions (i) and (ii) in Theorem 3.14, part (a). Firstly, note that, although it is possible to calculate *a priori* an upper bound on γ (see Doyle and Chu, 1986 for example), all such bounds tend to be excessively conservative or computationally expensive to evaluate. The procedure described here avoids such a calculation, and is as follows:

1. Select γ satisfying condition (i), i.e. $\gamma > \gamma_l$.
2. For this γ check that X_∞, Y_∞ satisfy condition (ii):
   - If 'YES', let $\gamma_u = \gamma$ and $\gamma = \frac{\gamma_l + \gamma_u}{2}$.
   - If 'NO', let $\gamma_l = \gamma$, $\gamma = 2\gamma$, and repeat this step.
3. For this γ check that X_∞, Y_∞ satisfy condition (ii):
   - If 'YES', let $\gamma_u = \gamma$ and $\gamma = \frac{\gamma_l + \gamma_u}{2}$ and if $\gamma_u - \gamma > \mu$ repeat this step.
   - If 'NO', let $\gamma_l = \gamma$ and $\gamma = \frac{\gamma_l + \gamma_u}{2}$ and repeat this step.

4. Let $\hat{\gamma}_{min} = \gamma_u$ be an approximation for γ_{min} and use this to solve for $\boldsymbol{K_a}$ (or $\tilde{\boldsymbol{K}}_a$).

By this method, the minimum achievable γ can be evaluated to within a designer selected tolerance μ.

Remark 3.17

In the case $m_2 = p_1$ and $p_2 = m_1$ (corresponding to $\boldsymbol{P}_{12}$ and $\boldsymbol{P}_{21}$ being square in the standard plant, $\boldsymbol{P}$) it is possible to find γ_{min} exactly, as is outlined, for example, in Limebeer and Hung, (1987). Hence the iterative procedure outlined above can be avoided. In the next chapter we will show that, in special cases, exact solutions for γ_{min} exist even if the above dimensional equalities do not hold.

3.5 Summary

This chapter has provided a generalized framework for the analysis and synthesis of robustly stabilizing controllers. In particular, we have shown that H_∞ control methods are suitable for robust controller synthesis, and that, if necessary, performance objectives can be incorporated into the same design.

CHAPTER FOUR

ROBUST STABILIZATION OF NORMALIZED COPRIME FACTOR PLANT DESCRIPTIONS

4.1 Introduction and Problem Statement

The previous chapter motivated the use of H_∞ optimization as a method for designing robustly stabilizing controllers, and a solution procedure for obtaining an optimal controller was outlined. In this chapter we are going to examine a specific robust stabilization problem, where we consider perturbations to the 'normalized left coprime factors' of a nominal plant, and select the set of stabilizing controllers which allow the H_∞ norm of the coprime factor perturbations to be as large as possible while maintaining internal stability. We will show that this robust stabilization problem has a particularly simple solution.

From Definition 2.19, a normalized left coprime factorization of a nominal plant G is a left coprime factorization $(\tilde{N}, \tilde{M})$ of G which satisfies

$$\tilde{N}\tilde{N}^* + \tilde{M}\tilde{M}^* = I \text{ for all } s \in j\Re, \tag{4.1}$$

or equivalently, $[\tilde{N}, \tilde{M}]$ is co-inner. We are therefore representing the nominal plant by a particular choice of coprime factors which, by Proposition 2.20, is unique to within left multiplication by a unitary matrix.

We will now use Corollary 3.7 (choosing the coprime factors to be normalized) to formally state the normalized LCF robust stabilization problem as an equivalent H_∞ optimization problem. Firstly though, we define the perturbation class associated with normalized coprime factor uncertainty: (as usual G, has dimensions $p \times m$)

$$\mathcal{D}_{C_\epsilon} \triangleq \{\Delta = [\Delta_N, \Delta_M];\ \Delta \in RH_\infty^{p\times(m+p)};\ \|\Delta\|_\infty < \epsilon\} \tag{4.2}$$

The normalized LCF uncertainty model is illustrated in Figure 4.1.

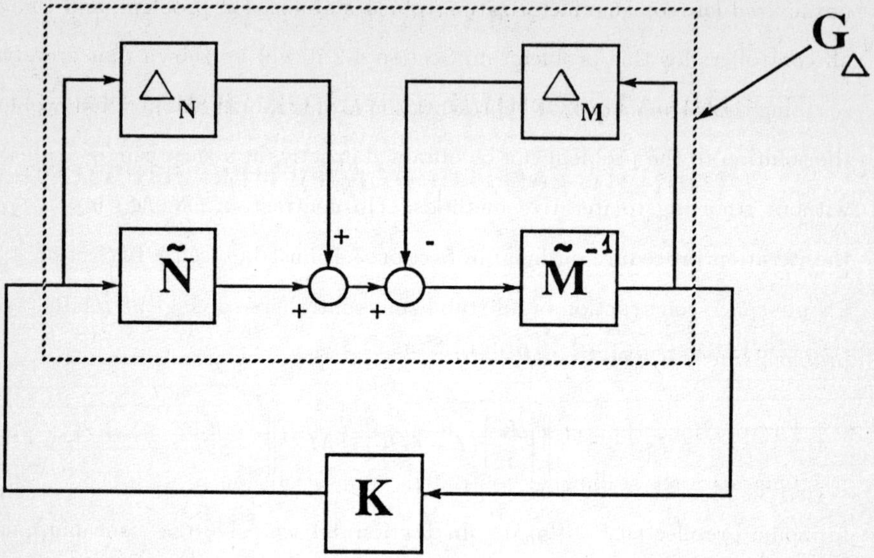

Figure 4.1 Coprime Factor Robust Stabilization Problem

Proposition 4.1 now states the normalized LCF robust stabilization problem :

Proposition 4.1 *The Normalized LCF Robust Stabilization Problem*

'Let $(\tilde{N}, \tilde{M})$ be a normalized left coprime factorization of G. Then, the largest positive number, ϵ $(= \epsilon_{max})$, such that $G_\Delta = (\tilde{M} + \Delta_M)^{-1}(\tilde{N} + \Delta_N)$ can be stabilized by a single controller, K, for all $\Delta \in \mathcal{D}_{C\epsilon}$ is given by

$$\epsilon_{max} = (\gamma_{min})^{-1} = \left(\inf_{K} \left\| \begin{bmatrix} K \\ I \end{bmatrix} (I - GK)^{-1} \tilde{M}^{-1} \right\|_\infty \right)^{-1} \tag{4.3}$$

where K is chosen over all stabilizing controllers.

Proof:

The proof is immediate from Corollary 3.7. ■

The remainder of this chapter is concerned with obtaining a solution for ϵ_{max} in this normalized left coprime factorization robust stabilization problem, and characterizing all controllers for this problem. In Section 4.2 it will be shown that an advantage of selecting the *normalized* LCF representation in this robust stabilization problem is that the solution to the problem can be obtained exactly, in a surprisingly explicit manner, without resorting to iterative methods. (In contrast, if $(\tilde{N}, \tilde{M})$ is *not* normalized, the iteration procedure outlined in Section 3.4 must be used.) In Section 4.3 we give a state-space construction of all stabilizing controllers solving the related *sub-optimal* problem where K must satisfy

$$\left\| \begin{bmatrix} K \\ I \end{bmatrix} (I - GK)^{-1} \tilde{M}^{-1} \right\|_\infty \leq \gamma$$

for some preselected $\gamma > \gamma_{min}$. In Section 4.4 we solve the same normalized LCF robust stabilization problem using the Glover and Doyle, (1988) method, and show clearly why no iteration on γ is needed to solve for γ_{min}. In Section 4.5 we examine a number of related issues: we show that a generalization of the normalized LCF robust stabilization problem leads to a class of related H_∞ problems with exact solutions; and we consider robust stabilization of the normalized LCF of positive real systems. Results are reviewed in Section 4.6.

4.2 Characterizing all Solutions

4.2.1 Solution Via a Nehari Extension Approach

We will now solve the normalized LCF robust stabilization problem stated in Proposition 4.1. An exact solution will be given using a method which presents a diversion from the general solution methods of Glover and Doyle given in Section 3.4. We will show that a coprime factorization of the optimally robust controller can be generated *directly* from the normalized coprime factors $(\tilde{N}, \tilde{M})$ of the nominal plant G by obtaining a *Nehari extension* of the matrix transfer function $\begin{bmatrix} -\tilde{N}^* \\ \tilde{M}^* \end{bmatrix}$. Briefly, an *optimal* Nehari extension problem involves approximating an unstable transfer function, R, by a stable transfer function, Q, where Q is chosen such that $\|R+Q\|_\infty$ is as small as possible. Glover, (1984) has shown that

$$\inf_{Q \in RH_\infty} \|R + Q\|_\infty = \|R^*\|_H. \tag{4.3}$$

In the related *sub-optimal* Nehari extension problem, Q is chosen such that

$$\|R + Q\|_\infty \leq \alpha,$$

where $\alpha > \|R^*\|_H$.

The main results of this section will now be stated, which relate the solution of the normalized LCF robust stabilization problem to the solution of a Nehari extension problem, and the optimal stability margin for this problem is explicitly stated.

Theorem 4.2

A controller, K, is stabilizing and satisfies

$$\left\| \begin{bmatrix} K \\ I \end{bmatrix} (I - GK)^{-1} \tilde{M}^{-1} \right\|_\infty \leq \gamma \tag{4.4}$$

if and only if $K = UV^{-1}$ for some $U, V \in RH_\infty$ satisfying

$$\left\| \begin{bmatrix} -\tilde{N}^* \\ \tilde{M}^* \end{bmatrix} + \begin{bmatrix} U \\ V \end{bmatrix} \right\|_\infty \leq (1 - \gamma^{-2})^{\frac{1}{2}}. \tag{4.5}$$

Theorem 4.3

(a) Optimal solutions to the normalized LCF robust stabilization problem give

$$\inf_K \left\| \begin{bmatrix} K \\ I \end{bmatrix} (I - GK)^{-1} \tilde{M}^{-1} \right\|_\infty = \{1 - \|[\tilde{N}, \tilde{M}]\|_H^2\}^{-1/2}$$

where the infimum is taken over all stabilizing K.

(b) The maximum stability margin is

$$\epsilon_{max} = \{1 - \|[\tilde{N}, \tilde{M}]\|_H{}^2\}^{1/2} > 0 \tag{4.6}$$

(c) All optimal controllers are given by $K = UV^{-1}$ *where* U, V, $\in RH_\infty$ *satisfy*

$$\left\| \begin{bmatrix} -\tilde{N}^* \\ \tilde{M}^* \end{bmatrix} + \begin{bmatrix} U \\ V \end{bmatrix} \right\|_\infty = \|[\tilde{N}, \tilde{M}]\|_H.$$

The proof of Theorem 4.2 will need the following Lemma. (The proof is given in Appendix C.)

Lemma 4.4

(a) Let E_1, $E_2 \in RL_\infty$ *satisfy*

$$\left\| \begin{bmatrix} E_1 \\ E_2 \end{bmatrix} \right\|_\infty \le \alpha < 1. \tag{4.7}$$

Then, $E_1(I - E_2)^{-1} \in RL_\infty$ *and*

$$\|E_1(I - E_2)^{-1}\|_\infty \le \alpha(1 - \alpha^2)^{-1/2}. \tag{4.8}$$

(b) For any $F \in RH_\infty$ *satisfying* $\|F\|_\infty \le \alpha(1 - \alpha^2)^{-1/2}$, $\alpha < 1$, *there exists* E_1, $E_2 \in RH_\infty$ *such that* $E_1(I - E_2)^{-1} = F$ *and (4.7) is satisfied.*

Proof of Theorem 4.2

(a) Necessity : suppose K is a stabilizing controller satisfying (4.4) and with RCF $(\hat{U}, \hat{V})$, then an equivalent RCF is (U, V) with

$$\begin{bmatrix} U \\ V \end{bmatrix} = -\gamma^{-2} \begin{bmatrix} \hat{U} \\ \hat{V} \end{bmatrix} (\tilde{M}\hat{V} - \tilde{N}\hat{U})^{-1} \tag{4.9}$$

since $(\tilde{M}\hat{V} - \tilde{N}\hat{U})^{-1} \in RH_\infty$ by internal stability requirements. Further, substituting for K and G in the closed-loop, and defining $S \triangleq (I - GK)^{-1}$, gives

$$\begin{bmatrix} K \\ I \end{bmatrix} (I - GK)^{-1} \tilde{M}^{-1} = \begin{bmatrix} \hat{U} \\ \hat{V} \end{bmatrix} (\tilde{M}\hat{V} - \tilde{N}\hat{U})^{-1} = -\gamma^2 \begin{bmatrix} U \\ V \end{bmatrix}$$

and as $\begin{bmatrix} M^* & N^* \\ -\tilde{N} & \tilde{M} \end{bmatrix}$ is all pass (where (N, M) is a *normalized* RCF of G satisfying $N^*N + M^*M = I$), we can exploit the unitary invariance of the H_∞ norm to obtain:

$$\gamma^2 \ge \left\| \begin{bmatrix} K \\ I \end{bmatrix} (I - GK)^{-1} \tilde{M}^{-1} \right\|_\infty^2$$

$$
\begin{aligned}
&= \left\| \begin{bmatrix} M^* & N^* \\ -\tilde{N} & \tilde{M} \end{bmatrix} \begin{bmatrix} \hat{U} \\ \hat{V} \end{bmatrix} (\tilde{M}\hat{V} - \tilde{N}\hat{U})^{-1} \right\|_\infty^2 \\
&= \left\| \begin{bmatrix} (M^*\hat{U} + N^*\hat{V})(\tilde{M}\hat{V} - \tilde{N}\hat{U})^{-1} \\ I \end{bmatrix} \right\|_\infty^2 \\
&= 1 + \|(M^*\hat{U} + N^*\hat{V})(\tilde{M}\hat{V} - \tilde{N}\hat{U})^{-1}\|_\infty^2 \qquad (4.10)
\end{aligned}
$$

Now consider

$$
\begin{aligned}
&\left\| \begin{bmatrix} -\tilde{N}^* \\ \tilde{M}^* \end{bmatrix} + \begin{bmatrix} U \\ V \end{bmatrix} \right\|_\infty^2 \\
&= \left\| \begin{bmatrix} M^* & N^* \\ -\tilde{N} & \tilde{M} \end{bmatrix} \left\{ \begin{bmatrix} -\tilde{N}^* \\ \tilde{M}^* \end{bmatrix} + \begin{bmatrix} U \\ V \end{bmatrix} \right\} \right\|_\infty^2 \\
&= \left\| \begin{bmatrix} -\gamma^{-2}(M^*\hat{U} + N^*\hat{V})(\tilde{M}\hat{V} - \tilde{N}\hat{U})^{-1} \\ (1-\gamma^{-2})I \end{bmatrix} \right\|_\infty^2 \quad \text{by (4.9)} \\
&\le (1-\gamma^{-2})^2 + \gamma^{-4}(\gamma^2 - 1) \quad \text{from (4.10)} \\
&= 1 - \gamma^{-2}
\end{aligned}
$$

and necessity is proven.

(b) Sufficiency : assume now that $U, V \in RH_\infty$ satisfies (4.5) and define

$$
\begin{aligned}
\begin{bmatrix} E_1 \\ E_2 \end{bmatrix} &= \begin{bmatrix} M^* & N^* \\ -\tilde{N} & \tilde{M} \end{bmatrix} \left\{ \begin{bmatrix} -\tilde{N}^* \\ \tilde{M}^* \end{bmatrix} + \begin{bmatrix} U \\ V \end{bmatrix} \right\} \\
&= \begin{bmatrix} M^*U + N^*V \\ I + (\tilde{M}V - \tilde{N}U) \end{bmatrix}
\end{aligned}
$$

where $\left\| \begin{bmatrix} E_1 \\ E_2 \end{bmatrix} \right\|_\infty^2 \le 1 - \gamma^{-2}$ by assumption. Further, for $K = UV^{-1}$;

$$
\begin{aligned}
\begin{bmatrix} M^* & N^* \\ -\tilde{N} & \tilde{M} \end{bmatrix} \begin{bmatrix} K \\ I \end{bmatrix} (I - GK)^{-1} \tilde{M}^{-1} &= \begin{bmatrix} M^* & N^* \\ -\tilde{N} & \tilde{M} \end{bmatrix} \begin{bmatrix} U \\ V \end{bmatrix} (\tilde{M}V - \tilde{N}U)^{-1} \\
&= \begin{bmatrix} E_1(E_2 - I)^{-1} \\ I \end{bmatrix}
\end{aligned}
$$

and applying Lemma 4.4(a)

$$
\left\| \begin{bmatrix} K \\ I \end{bmatrix} (I - GK)^{-1} \tilde{M}^{-1} \right\|_\infty^2 \le 1 + \frac{(1-\gamma^{-2})}{\gamma^{-2}} = \gamma^2
$$

Finally, K will be stabilizing, and (U, V) is an RCF of K, since $(\tilde{M}V - \tilde{N}U)^{-1} = (E_2 - I)^{-1} \in RH_\infty$ by the Small Gain Theorem since $\|E_2\|_\infty < 1$ by assumption and $E_2 \in RH_\infty$. ∎

To prove Theorem 4.3 we require a further technical Lemma. (The proof is again given in Appendix C.)

Lemma 4.5

Let $(\tilde{N}, \tilde{M})$ be a normalized LCF. Then

$$\|[\tilde{N}, \tilde{M}]\|_H < 1.$$

Proof of Theorem 4.3:

This is an immediate consequence of Theorem 4.2, Lemma 4.5 and Nehari's Theorem, which, by (4.3), states that

$$\inf_{U,V \in RH_\infty} \left\| \begin{bmatrix} -\tilde{N}^* \\ \tilde{M}^* \end{bmatrix} + \begin{bmatrix} U \\ V \end{bmatrix} \right\|_\infty = \|[\tilde{N}, \tilde{M}]\|_H.$$

Then, by (4.5) and Nehari's Theorem, $1-\gamma^{-2} \geq \|[\tilde{N}, \tilde{M}]\|_H^2$, with equality achievable by choosing $\begin{bmatrix} U \\ V \end{bmatrix}$ to be an optimal Nehari extension of $\begin{bmatrix} -\tilde{N}^* \\ \tilde{M}^* \end{bmatrix}$. Lemma 4.5 guarantees that $\epsilon_{max} > 0$. ■

Remark 4.6

The main implication of Theorems 4.2 and 4.3 is that all H_∞ optimal controllers for this problem can be obtained by solving a related Nehari extension problem. Theoretically, this is a more straightforward approach than a general H_∞ approach, and it is hoped that this will give greater insight into the nature of this problem.

Remark 4.7

The maximum stability margin given in (4.6) is surprisingly explicit. The H_∞ optimization problem posed in (4.2) is a problem for which an iterative solution is normally required to approximate ϵ_{max}. (See Section 3.4, Remark 3.16.) When we examine the approach of Glover and Doyle, (1988) in Section 4.4, we will show that, for the normalized LCF robust stabilization problem, the suggested iteration procedure can be dispensed with.

Remark 4.8

It is instructive at this point to examine the implication of a small stability margin– that is $\epsilon_{max} \ll 1$. Noting that, $G_\Delta = (\tilde{M} + \Delta_M)^{-1}(\tilde{N} + \Delta_N)$, then in the gain cross-over frequency range, both $\overline{\sigma}(\tilde{N})$ and $\overline{\sigma}(\tilde{M})$ are of order one, and hence obtaining $\epsilon_{max} \ll 1$ in this frequency range implies that there exists a small proportional perturbation on $\tilde{M}$ or $\tilde{N}$ which will destabilize the system. However at frequencies of high gain ($\underline{\sigma}(G) \gg 1$) we have $\overline{\sigma}(\tilde{M}) \ll 1$, and even if $\epsilon_{max} \ll 1$ this may still imply that a *large* proportional perturbation on $\tilde{M}$ is required to destabilize the closed-loop system. Similarly, at frequencies of low gain ($\overline{\sigma}(G) \ll 1$) we have $\overline{\sigma}(\tilde{N}) \ll 1$, and even if $\epsilon_{max} \ll 1$ this may still imply that a *large* proportional perturbation on $\tilde{N}$ is required to destabilize the closed-loop system.

Remark 4.9

It is possible to obtain a dual result for the maximum stability margin if a normalized *right* coprime factorization (N, M) of the nominal plant is used. The perturbed plant would be

$$G_\Delta = (N + \Delta_N)(M + \Delta_M)^{-1} \tag{4.11}$$

where N, M are normalized. The maximum stability margin for this problem can be shown to be

$$\epsilon_{max} = (1 - \left\| \begin{bmatrix} N \\ M \end{bmatrix} \right\|_H^2)^{1/2}$$

which implies that stability of the closed-loop can be guaranteed if

$$\left\| \begin{bmatrix} \Delta_N \\ \Delta_M \end{bmatrix} \right\|_\infty < \epsilon_{max}$$

It is interesting to note that for any nominal plant, the optimal stability margin for the normalized left and right coprime factor problems is the same. It is also obvious that all optimal controllers for the normalized RCF problem can be obtained from

$\tilde{U}, \tilde{V} \in RH_\infty$ such that $K = \tilde{V}^{-1}\tilde{U}$, where $[\tilde{U}, \tilde{V}]$ is the optimal Nehari extension of $[-N^*, M^*]$.

Remark 4.10

In the robust stabilization of $(G+\Delta)$ with $\|\Delta\|_\infty < \beta$ it has been observed (Glover, 1986a) that the largest robustly stabilizable region with a single controller has a "non-stabilizable" plant on its boundary. This result has been exploited by Khargonekar, Georgiou and Pascoal, (1987) to show, for example, that non-linear and time-varying controllers can do no better. The same comments apply to unstructured multiplicative perturbations. However, in the case of coprime factor uncertainty, it can be shown that, in general, the *nearest* unstabilizable system[1] to a nominal $[\tilde{N}, \tilde{M}]$ (in the H_∞ norm) is beyond the boundary of the largest robustly stabilizable set obtainable using a single linear, time invariant, controller: Consider the following second order example. A nominal plant $G = 12/s(s+5)$ has a normalized left coprime factorization given by

$$[\tilde{N}, \tilde{M}] = [12, s(s+5)]/(s+3)(s+4)$$

Routine calculations yield that $\|[\tilde{N}, \tilde{M}]\|_H = 0.79156$, and hence $\epsilon_{max} = 0.61109$. However, in the single-input, single-output case, the distance to the nearest stabilizable plant is given by

$$d = \inf_{Re(s)>0} (|\tilde{N}|^2 + |\tilde{M}|^2)^{1/2}$$

where d indicates the minimum euclidean distance to a RHP pole/zero cancellation. A search on $Re(s) > 0$ indicates that $d = 0.6146$, and thus the nearest unstabilizable system to $[\tilde{N}, \tilde{M}]$ in this norm is very slightly beyond the boundary of the largest robustly stabilizable set. Hence the above remarks concerning non-linear and time-varying controllers do not necessarily apply in this case.

---

1 By *nearest unstabilizable system*, we mean here the nearest $G_\Delta = (N+\Delta_N)(M+\Delta_M)^{-1}$ such that the pair $(N+\Delta_N, M+\Delta_M)$ is no longer coprime. That is, there is RHP pole/zero cancellation.

4.2.2 Parametrizing All Controllers

A related problem to the optimal H_∞ problem posed in Proposition 4.1, is the sub-optimal problem, of obtaining the set of stabilizing controllers K such that

$$\left\| \begin{bmatrix} K \\ I \end{bmatrix} (I - GK)^{-1} \tilde{M}^{-1} \right\|_\infty \leq \gamma \tag{4.12}$$

where $\gamma(> \epsilon_{max}^{-1})$ is some prespecified tolerance level for the allowable uncertainty. Theorem 4.2 shows that this is equivalent to finding all stable extensions of $\begin{bmatrix} -\tilde{N}^* \\ \tilde{M}^* \end{bmatrix}$, designated Q, such that

$$\left\| \begin{bmatrix} -\tilde{N}^* \\ \tilde{M}^* \end{bmatrix} + Q \right\|_\infty \leq \alpha \tag{4.13}$$

where $\alpha = (1 - \gamma^{-2})^{1/2}$.

Appendix B gives a characterization of all Q satisfying (4.13), and it is thus possible to characterize all controllers achieving (4.12).

Theorem 4.11

For $1 > \alpha > \|[\tilde{N}, \tilde{M}]\|_H$, let the parametrization of all extensions of $\begin{bmatrix} -\tilde{N}^* \\ \tilde{M}^* \end{bmatrix}$ such that $Q \in RH_\infty$ and

$$\left\| \begin{bmatrix} -\tilde{N}^* \\ \tilde{M}^* \end{bmatrix} + Q \right\|_\infty \leq \alpha$$

be given by (see Appendix B)

$$Q = \mathcal{T}_J[\Phi],$$

$\Phi \in RH_\infty^{(p+m)\times p}$, $\|\Phi\|_\infty \leq 1$, *and* $J \in RH_\infty^{((p+m)+p)\times(m+p+p)}$,

$$J = \left[\begin{array}{cc|c} J_{11} & -\alpha J_{12} & J_{12} \\ \hline J_{21} & I - \alpha J_{22} & J_{22} \end{array} \right]. \tag{4.14}$$

Then, all stabilizing controllers satisfying (4.12) are given by

$$K = (\tilde{J}_{11u}\tilde{\Phi} + \tilde{J}_{12u})(\tilde{J}_{11v}\tilde{\Phi} + \tilde{J}_{12v})^{-1}$$

where $J_{11} = \begin{bmatrix} J_{11u} \\ J_{11v} \end{bmatrix}$, $J_{12} = \begin{bmatrix} J_{12u} \\ J_{12v} \end{bmatrix}$ *are partitioned conformally with* $\begin{bmatrix} -\tilde{N}^* \\ \tilde{M}^* \end{bmatrix}$ *and*

$$\begin{bmatrix} \tilde{J}_{11u} & \tilde{J}_{12u} \\ \tilde{J}_{11v} & \tilde{J}_{12v} \end{bmatrix} \triangleq \begin{bmatrix} \gamma J_{11u} & J_{12u} \\ \gamma J_{11v} & J_{12v} \end{bmatrix}$$

and $\tilde{\Phi} \in RH_\infty^{m\times p}$ *such that* $\|\tilde{\Phi}\|_\infty \leq 1$.

Proof:

From (4.14), all Q are given by

$$Q = (J_{11}\Phi_1 - \alpha J_{12}\Phi_2 + J_{12})(J_{21}\Phi_1 + (I - \alpha J_{22})\Phi_2 + J_{22})^{-1}$$

where $\Phi = \begin{bmatrix} \Phi_1 \\ \Phi_2 \end{bmatrix}$, and $\Phi_1 \in RH_\infty^{m\times p}, \Phi_2 \in RH_\infty^{p\times p}$. Noting that, as $Q \in RH_\infty$ and $J \in RH_\infty$ it follows that $(J_{21}\Phi_1 + (I - \alpha J_{22})\Phi_2 + J_{22})$ is a unit in RH_∞, then from basic coprime factorization theory (see Lemma 2.18) coprime factorizations of all controllers (given by Q), are also given by

$$\begin{aligned} Q' &= (J_{11}\Phi_1 - \alpha J_{12}\Phi_2 + J_{12}) \\ &= J_{11}\Phi_1 + J_{12}(I - \alpha\Phi_2) \end{aligned}$$

where $(I - \alpha\Phi_2) \in RH_\infty$, and also $(I - \alpha\Phi_2)^{-1} \in RH_\infty$ by the Small Gain Theorem. Hence without loss of generality, coprime factorizations of all controllers are given by

$$Q'' = J_{11}\Phi_1(I - \alpha\Phi_2)^{-1} + J_{12}.$$

Next, noting that

$$\left\| \begin{bmatrix} \alpha\Phi_1 \\ \alpha\Phi_2 \end{bmatrix} \right\|_\infty \leq \alpha < 1$$

then by Lemma 4.4(a)

$$\Rightarrow \|\tilde{\Phi}_1\|_\infty \leq (1 - \alpha^2)^{-1/2}$$

where $\tilde{\Phi}_1 \triangleq \Phi_1(I - \alpha\Phi_2)^{-1}$. Further, by Lemma 4.4(b), the converse is also always true for some Φ_1, Φ_2. Hence, coprime factors of all controllers are now given by

$$Q'' = J_{11}\tilde{\Phi}_1 + J_{12},$$

where $\tilde{\Phi}_1 \in RH_\infty^{m\times p}$ and $\|\tilde{\Phi}_1\|_\infty \leq (1 - \alpha^2)^{-1/2}(= \gamma)$. Alternatively, by scaling $\tilde{\Phi}_1$ by γ^{-1} and J_{11} by γ, we have

$$Q'' = \tilde{J}_{11}\tilde{\Phi} + \tilde{J}_{12},$$

where $\tilde{J}_{11} = \gamma J_{11}$, $\tilde{J}_{12} = J_{12}$, $\tilde{\Phi} \in RH_\infty^{m\times p}$ and $\|\tilde{\Phi}\|_\infty \leq 1$

Noting that $Q'' = \begin{bmatrix} U \\ V \end{bmatrix}$, where U, V are coprime factors of K: $K = UV^{-1}$, then appropriate partitioning of J_{11}, J_{12} as $\begin{bmatrix} J_{11u} \\ J_{11v} \end{bmatrix}$, $\begin{bmatrix} J_{12u} \\ J_{12v} \end{bmatrix}$ yields the desired result.

■

Remark 4.12

It is interesting to note that although there are $(p+m)\times p$ degrees of freedom associated with Q (via the arbitrary contraction Φ), there are only $m \times p$ degrees of freedom associated with the resulting sub-optimal controller (via $\tilde{\Phi}$). The reason for this is that the sub-optimal extension can generate right coprime factorizations by multiplying any one solution by a $p \times p$ unit in RH_∞. The remaining $m \times p$ degrees of freedom are associated with the controller (rather than the factorization) and would also be obtained had the controller been derived via the H_∞ methods of Glover and Doyle, (1988). (See Section 4.4.)

In the next section, a state-space characterization of all sub-optimal controllers will be given.

4.3 A State Space Representation of Sub-Optimal Controllers

4.3.1 Characterizing All Controllers

We now give a state-space characterization of all sub-optimal controllers for the normalized LCF robust stabilization problem – that is, all stabilizing K such that

$$\left\| \begin{bmatrix} K \\ I \end{bmatrix} (I - GK)^{-1} \tilde{M}^{-1} \right\|_\infty \leq \gamma. \tag{4.15}$$

Firstly, recall from Proposition 2.22 that a state-space realization of $[\tilde{N}, \tilde{M}]$ is given by

$$[\tilde{N}, \tilde{M}] \stackrel{s}{=} \left[\begin{array}{c|cc} A^o & B + HD & H \\ \hline R^{-1/2}C & R^{-1/2}D & R^{-1/2} \end{array} \right], \tag{4.16}$$

where $A^o = A + HC$; $H = -(ZC^* + BD^*)R^{-1}$, and $R = I + DD^*$. We now state a preliminary result showing that the controllability and observability gramians of $[\tilde{N}, \tilde{M}]$ can be related to the solutions of GCARE and GFARE.

Lemma 4.13

The controllability and observability gramians of $[\tilde{N}, \tilde{M}]$, P *and* Q *respectively, are given by*

$$P = Z$$

$$Q = X(I + ZX)^{-1}$$

where X, Z *are the unique positive definite solutions to GCARE and GFARE respectively.*

Proof:

The proof is simple by noting Definition 2.1, the realization in (4.16), and the properties of the solutions to GCARE and GFARE given in Appendix A, where P follows from (A.16), and Q from (A.18). ∎

From (4.16), a minimal state-space representation of $\begin{bmatrix} -\tilde{N}^* \\ \tilde{M}^* \end{bmatrix}$ is

$$\left[\begin{array}{c|c} -(A^o)^* & C^* R^{-1/2} \\ \hline (B + HD)^* & -D^* R^{-1/2} \\ -H^* & R^{-1/2} \end{array}\right] \tag{4.17}$$

and by Lemma 4.13, solutions to the Lyapunov equations associated with this realization of $\begin{bmatrix} -\tilde{N}^* \\ \tilde{M}^* \end{bmatrix}$ are given by:

$$P = -X(I + ZX)^{-1} \tag{4.18}$$

and

$$Q = -Z \tag{4.19}$$

The results of Theorem 4.11 combined with Lemma B.2 (Appendix B) yield the following state-space parametrization for all sub-optimal controllers satisfying (4.15).

Theorem 4.14

All controllers for the normalized LCF robust stabilization problem satisfying $\|\mathcal{F}(P,K)\|_\infty \le \gamma$*, for* $\gamma > \gamma_{min} = (1 - \|[\tilde{N},\tilde{M}]\|_H^2)^{-1/2}$*, are given by*

$$K = \mathcal{T}_L[\Phi] \tag{4.20}$$

where L *has the state-space form*

$$L = \left[\begin{array}{c:c} L_{11} & L_{12} \\ \hdashline L_{21} & L_{22} \end{array}\right] = \left[\begin{array}{c|c:c} A^c & -\gamma^2 W_1^{*-1} B S^{-1/2} & \gamma^2 \zeta^{-1} W_1^{*-1} Z C^* R^{-1/2} \\ \hline F & S^{-1/2} & \zeta^{-1} D^* R^{-1/2} \\ \hdashline C + DF & DS^{-1/2} & -\zeta^{-1} R^{-1/2} \end{array}\right] \tag{4.21}$$

where $\zeta = (\gamma^2 - 1)^{1/2}$,

$$W_1 \triangleq I + (XZ - \gamma^2 I), \tag{4.22}$$

$\Phi \in RH_\infty^{m\times p}$ *with* $\|\Phi\|_\infty \le 1$*,* $G \stackrel{s}{=} (A,B,C,D)$ *is the nominal plant,* X*,* Z *solve GCARE, GFARE,* $S = I + D^*D$*,* $F = -S^{-1}(B^*X + D^*C)$ *and* $A^c = A + BF$*.*

Proof:

As $[\tilde{N},\tilde{M}]$ is a co-inner function, the sub-optimal extension of $\begin{bmatrix} -\tilde{N}^* \\ \tilde{M}^* \end{bmatrix}$ will have the form given in (B.6) in Appendix B. Noting from Theorem 4.11 that only the (1,1) and (1,2) blocks of this state-space expression are required to determine all controllers, the state-space form of L (as required for (4.20)) can immediately be written as:

$$L = \left[\begin{array}{c|c} A_L & B_L \\ \hline C_L & D_L \end{array}\right]$$

and by noting that $D_\perp = \begin{bmatrix} S^{-1/2} \\ DS^{-1/2} \end{bmatrix}$ is a suitable unitary completion of $\begin{bmatrix} D^*R^{-1/2} \\ -R^{-1/2} \end{bmatrix}$ then :

$$A_L = A^o \tag{4.23}$$

$$B_L = [\, W^{*-1} B S^{-1/2} \quad \alpha^{-1} W^{*-1} Q C^* R^{-1/2} \,] \tag{4.24}$$

$$C_L = \left[-\begin{pmatrix} (B+HD)^* \\ -H^* \end{pmatrix} P + \begin{pmatrix} DR^{-1/2} \\ -R^{-1/2} \end{pmatrix} R^{-1/2} C \right] \tag{4.25}$$

$$D_L = \begin{bmatrix} S^{-1/2} & \alpha^{-1} D^* R^{-1/2} \\ DS^{-1/2} & -\alpha^{-1} R^{-1/2} \end{bmatrix} \tag{4.26}$$

where the (2,1) and (2,2) blocks of the parametrization in (B.6) have been omitted here, and α is the tolerance for the sub-optimal extension in (4.13). In addition

$$W \triangleq PQ - \alpha^2 I$$

and P and Q are defined in (4.18) and (4.19).

Now, by noting that

$$\begin{aligned} W &= XZ(I+XZ)^{-1} - \alpha^2 I \\ &= (\gamma^{-2}(I+XZ) - I)(I+XZ)^{-1} \text{ as } \alpha^2 = 1-\gamma^{-2} \\ \Rightarrow W^{*-1} &= \gamma^2 (I+ZX) W_1^{*-1} \end{aligned}$$

where $W_1 \triangleq I + (XZ - \gamma^2 I)$.

Then, (4.24) can be written:

$$B_L = \gamma^2(I+ZX)W_1^{*-1}[BS^{-1/2}, -\alpha^{-1}ZC^*R^{-1/2}].$$

Next note that

$$\begin{aligned} H^*P - R^{-1}C &= -H^*X(I+ZX)^{-1} - R^{-1}C \\ &= R^{-1}(DB^*X - C)(I+ZX)^{-1} \text{ by (A.10)} \\ &= -(C+DF)(I+ZX)^{-1} \text{ by (A.5), (A.6) and (A.9),} \end{aligned}$$

and similarly

$$\begin{aligned} BP + D^*(H^*P - R^{-1}C) &= (B^*X + D^*(C+DF))(I+ZX)^{-1} \\ &= -F(I+ZX)^{-1} \text{ by (A.9).} \end{aligned}$$

Hence (4.25) can be written:

$$C_L = \begin{bmatrix} -F \\ -(C+DF) \end{bmatrix} (I+ZX)^{-1}.$$

Applying a state transformation of the form $-(I+ZX)^{-1}$ to $\boldsymbol{L}$, and by noting (A.13), we have

$$\boldsymbol{L} = \left[\begin{array}{c:c} L_{11} & L_{12} \\ \hdashline L_{21} & L_{22} \end{array}\right] = \left[\begin{array}{c|c:c} A^c & -\gamma^2 W_1^{*-1} B S^{-1/2} & \gamma^2 \alpha^{-1} W_1^{*-1} Z C^* R^{-1/2} \\ \hline F & S^{-1/2} & \alpha^{-1} D^* R^{-1/2} \\ \hdashline (C+DF) & DS^{-1/2} & -\alpha^{-1} R^{-1/2} \end{array}\right]$$

where by Theorem 4.11, all controllers satisfying (4.15) are given by $K = \mathcal{T}_L[\psi]$ and $\|\psi\|_\infty \leq \gamma$. Finally, scaling of L_{11}, L_{21}, as in the proof of Theorem 4.11, and then noting that $\gamma^{-1}L$ also gives a parametrization of all sub-optimal controllers (note that $\alpha^{-1} = \gamma(\gamma^2 - 1)^{-1/2}$) we have the required result given in (4.21). ■

4.3.2 The Central Controller

A particular controller satisfying (4.15) is given by

$$K_o = L_{12}L_{22}{}^{-1},$$

corresponding to $\Phi = 0$, which is the *central* or *maximum entropy* controller (Glover and Mustafa, 1988) for the selected tolerance level. The state-space formulae for K_o is now given.

Corollary 4.15

The central controller achieving (4.15), has the state-space form

$$K_o = \left[\begin{array}{c|c} A^c + \gamma^2 W_1^{*-1} Z C^*(C + DF) & \gamma^2 W_1^{*-1} Z C^* \\ \hline B^* X & -D^* \end{array}\right] \tag{4.27}$$

where X (resp. Z) solves GCARE (resp. GFARE), and

$$F = -S^{-1}(D^*C + B^*X)$$

$$A^c = A + BF$$

$$W_1 = I + (XZ - \gamma^2 I).$$

Proof:

Figure 4.2 demonstrates that $K_o = L_{12}L_{22}{}^{-1}$ can be written as a unity feedback system. The nth order state-space form of K_o can be readily constructed by noting this feedback arrangement and using the realization in (4.21). (The construction is straightforward state-space manipulation.) ■

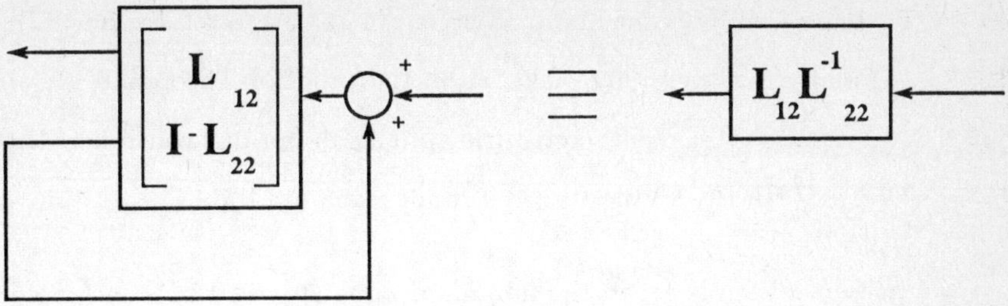

Figure 4.2 Construction of K from Coprime Factors

Remark 4.16

As has been recently stated by Doyle *et al*, (1988), the controller for the sub-optimal H_∞ problem requires only the solution to two Riccati equations (as is required for an LQG controller). The present problem requires one Riccati equation to obtain $[\tilde{N}, \tilde{M}]$, and thence pose a standard H_∞ problem. Thereafter one more Riccati equation need be solved as shown in Corollary 4.15. Further, noting from Lemma 4.13 that

$$\|[\tilde{N}, \tilde{M}]\|_H^2 = \lambda_{max}(PQ) = \lambda_{max}(ZX(I + ZX)^{-1}) \tag{4.28}$$

then by Lemma 4.13, and (4.6) it can be seen that (after some manipulation) :

$$\gamma_{min}^2 = \epsilon_{max}^{-2} = 1 + \lambda_{max}(ZX) \tag{4.29}$$

Hence, for the normalized LCF robust stabilization problem, the Riccati solutions are sufficient to specify the maximum achievable stability margin. Note that $\lambda_i(ZX)$ are precisely the closed-loop characteristic values defined in Jonckheere and Silverman, (1983).

Remark 4.17 - *Optimal Controllers*

If the tolerance level is set to the minimum value specified in (4.29), a controller of degree $\leq n-1$ exists. (See Limebeer and Halikias, 1988.) For this, the *optimal* case, it can be shown (Safonov, Chiang, and Limebeer, 1987) that as $\gamma \rightarrow \gamma_{min}$ only $n-r$ states will exist in the realization in (4.27), where $r \geq 1$ is the multiplicity of $\lambda_{max}(ZX)$. All such optimal controllers can be constructed from the optimal Nehari extension of $\begin{bmatrix} -\tilde{N}^* \\ \tilde{M}^* \end{bmatrix}$ (see Theorem 4.3) following the state-space construction in Glover, (1984).

Remark 4.18

When $D \neq 0$ the feedback system is always well-posed since $det(I + DD^*) \neq 0$. However if we now require that the system is well-posed in the face of infinitesimal time delays (see Willems, 1971), the condition $\overline{\sigma}(D) < 1$ is sufficient.

In the next section we will solve the normalized LCF robust stabilization problem using the general method of Glover and Doyle, (1988) and show that an identical maximum stability margin to (4.29), and hence (4.6), can be obtained.

4.4 Solution Using the Procedure of Glover and Doyle, (1988)

The results in Section 4.2 show that the normalized LCF robust stabilization problem can be solved exactly. The procedure presented represented a diversion from the more general solution approach outlined in Section 3.4, and the objective of this section is to show that the Glover and Doyle, (1988) approach *also* yields an exact solution to this problem despite its iterative appearance.

We now follow the procedure in Section 3.4, and first establish conditions for the existence of a stabilizing controller, K, such that

$$\left\| \begin{bmatrix} K \\ I \end{bmatrix} (I - GK)^{-1} \tilde{M}^{-1} \right\|_\infty \leq \gamma \tag{4.30}$$

as in Theorem 3.14(a). For simplicity, we will assume here that the nominal plant, G, is strictly proper, and has a state-space realization $G \stackrel{s}{=} (A, B, C, 0)$. Recalling from Section 3.3 that the Standard Plant, P, for the coprime factor uncertainty problem is given by

$$P = \left[\begin{array}{c:c} P_{11} & P_{12} \\ \hdashline P_{21} & P_{22} \end{array}\right] = \left[\begin{array}{c:c} \tilde{M}^{-1} & G \\ 0 & I \\ \hdashline \tilde{M}^{-1} & G \end{array}\right]$$

then an equivalent state-space form is (refer to Proposition 2.22)

$$P \stackrel{s}{=} \left[\begin{array}{c|c:c} A & B_1 & B_2 \\ \hline C_1 & D_{11} & D_{12} \\ \hdashline C_2 & D_{21} & D_{22} \end{array}\right] = \left[\begin{array}{c|c:c} A & -H & B \\ \hline C & I & 0 \\ 0 & 0 & I \\ \hdashline C & I & 0 \end{array}\right] \tag{4.31}$$

where $H = -ZC^*$, and Z solves the GFARE associated with G.

Assumptions (A1)-(A6) all hold provided $(A, B, C, 0)$ is a minimal realization, and we can use the procedure from Glover and Doyle, (1988). Appropriate substitution of (4.31) into (3.13) and (3.14) shows that the ARE solutions X_∞, Y_∞ must satisfy:

$$X_\infty(A + (1-\gamma^2)^{-1}HC) + (A + (1-\gamma^2)^{-1}HC)^* X_\infty$$

$$- X_\infty((1-\gamma^2)^{-1}HH^* + BB^*)X_\infty - \gamma^2(1-\gamma^2)^{-1}C^*C = 0 \tag{4.32}$$

$$Y_\infty(A + HC)^* + (A + HC)Y_\infty - Y_\infty C^* C Y_\infty = 0 \tag{4.33}$$

The appropriate X_∞ and Y_∞ solutions can be written

$$X_\infty = -\gamma^2 W_1^{-1} X \tag{4.34}$$

$$Y_\infty = 0 \tag{4.35}$$

where $W_1 \triangleq I + XZ - \gamma^2 I$, the matrices X, Z solve GCARE and GFARE respectively, and γ is the required tolerance level. To show that X_∞ solves (4.32), note that the LHS of (4.32) can be rewritten as

$$X_\infty A + A^* X_\infty - X_\infty BB^* X_\infty + C^*C - (1-\gamma^2)^{-1}(I + X_\infty Z)C^*C(I + ZX_\infty).$$

Then, using

$$X_\infty = -\gamma^2 W_1^{-1} X \Rightarrow I + X_\infty Z = W_1^{-1}(1-\gamma^2)(I + XZ) \tag{4.36}$$

and applying results (A.3) and (A.13) in Appendix A, the result is obtained. $Y_\infty = 0$ obviously solves (4.33).

To show that X_∞ and Y_∞ in (4.34) and (4.35) are the stabilizing solutions of the respective AREs, we require that

$$Re[\lambda_i(A + (1-\gamma^2)^{-1}HC - ((1-\gamma^2)^{-1}HH^* + BB^*)X_\infty)] < 0 \tag{4.37}$$

$$Re[\lambda_i(A + HC - Y_\infty C^* C)] < 0. \tag{4.38}$$

(4.38) is obviously satisfied, and (4.37) is also satisfied by noting

$$\begin{aligned}
&\lambda_i(A + (1-\gamma^2)^{-1}HC - ((1-\gamma^2)^{-1}HH^* + BB^*)X_\infty) \\
&= \lambda_i(A + HC(I + ZX)W_1^{*-1} + \gamma^2 BB^* X W_1^{*-1}) \text{ by (4.34), (4.36)} \\
&= \lambda_i(((A + HC)(I + ZX) - \gamma^2(A + BF))W_1^{*-1}) \\
&= \lambda_i(W_1^*(A + BF)W_1^{*-1}) = \lambda_i(A + BF) \text{ by (A.13).}
\end{aligned}$$

The corresponding gain matrices F_I and H_I are then

$$F_I = \begin{bmatrix} F_{12} \\ F_2 \end{bmatrix} = \begin{bmatrix} -C(I + ZX) \\ \gamma^2 B^* X W_1^{*-1} \end{bmatrix}$$

$$H_I = [\,H_{11} \quad H_{12} \quad H_2\,] = [\,0 \quad 0 \quad H\,].$$

We now apply conditions (i) and (ii) in Theorem 3.14(a), to obtain the admissible set of tolerances γ:

Condition (i) simply states:

$$\gamma > 1. \tag{4.39}$$

Condition (ii) can be simplified by noting that $Y_\infty = 0$ and hence $\rho(X_\infty Y_\infty) = 0$ where ρ denotes spectral radius. Further, $X_\infty \geq 0$ is satisfied if

$$-\gamma^2 W_1^{-1} X \geq 0$$

$$\Rightarrow \gamma^2 (I + XZ - \gamma^2 I)^{-1} X \leq 0$$

and noting that $X > 0$, we obtain the condition

$$\gamma^2 \geq 1 + \lambda_{max}(XZ) \tag{4.40}$$

Comparing (4.39) and (4.40) it is clear that the minimum allowable γ is

$$\gamma_{min}^2 = 1 + \lambda_{max}(XZ)$$

which is precisely the result obtained in (4.29).

Using Theorem 3.14(b) all stabilizing controllers achieving the tolerance γ in (4.30) can be parametrized for $\gamma > \gamma_{min}$. The selection $\Phi = 0$, and evaluation of equations (3.7)-(3.16) yields the 'central controller' for this problem, which has state-space form:

$$K \stackrel{s}{=} \left[\begin{array}{c|c} \hat{A} & \hat{B}_1 \\ \hline \hat{C}_1 & \hat{D}_{11} \end{array}\right] = \left[\begin{array}{c|c} A + HC + \gamma^2 BB^* X W_1^{*-1} & -H \\ \hline \gamma^2 B^* X W_1^{*-1} & 0 \end{array}\right] \tag{4.41}$$

Now, applying a state similarity transformation of the form $T = \gamma^2 W_1^{*-1}$ to the realization in (4.41), and using (A.13) to simplify the result we obtain an equivalent realization:

$$K \stackrel{s}{=} \left[\begin{array}{c|c} A + BF + \gamma^2 W_1^{*-1} Z C^* C & \gamma^2 W_1^{*-1} Z C^* \\ \hline B^* X & 0 \end{array}\right]$$

which is *exactly* the controller realization given in Section 4.3, equation (4.27) when $D = 0$.

Remark 4.19

It is not surprising that we obtain identical results here to those obtained in Sections 4.2 and 4.3, but it has been interesting to note how the iterative procedure for estimating γ_{min} was avoided. This was possible because it was possible to write ARE solutions X_∞ and Y_∞ *explicitly* in terms of γ, and the solutions to GCARE and GFARE (which are independent of γ), and hence the conditions of Theorem 3.14(a) could be written as explicit conditions on γ.

From a computational, as well as a theoretical point of view, it is obviously highly desirable to have H_∞ optimization problems with exact solutions. In the next section we examine a number of issues related to the normalized LCF robust stabilization problem and firstly, we show that exact solutions can be obtained for several other H_∞ problems.

4.5 Related Issues

4.5.1 H_∞ Problems with Exact Solutions

So far in this chapter we have shown that an exact solution can be derived for the normalized LCF robust stabilization problem and hence a computationally expensive iterative procedure can be bypassed. In this section we show that by slightly generalizing the H_∞ problem of Proposition 4.1, we can obtain exact solutions to a class of related H_∞ problems.

We begin by manipulating the results of Theorem 4.3 to give a preliminary result:

Lemma 4.20

Let $(\tilde{N}, \tilde{M})$ be a normalized left coprime factorization of G. Then

$$\inf_{K \text{ stabilizing}} \left\| \begin{bmatrix} I \\ K \end{bmatrix} (I - GK)^{-1} [I, G] \right\|_\infty = (1 - \|[\tilde{N}, \tilde{M}]\|_H^2)^{-1/2}. \tag{4.42}$$

Proof:

Note that a property of the H_∞ norm is that it is invariant under right (resp. left) multiplication by a co-inner (resp. inner) function. Hence recalling from (4.1) that $[\tilde{N}, \tilde{M}]$ is a co-inner function, we have that

$$\begin{aligned}\left\|\begin{bmatrix} K \\ I \end{bmatrix}(I-GK)^{-1}\tilde{M}^{-1}\right\|_\infty &= \left\|\begin{bmatrix} I \\ K \end{bmatrix}(I-GK)^{-1}\tilde{M}^{-1}[\tilde{M}, \tilde{N}]\right\|_\infty \\ &= \left\|\begin{bmatrix} I \\ K \end{bmatrix}(I-GK)^{-1}[I, G]\right\|_\infty\end{aligned}$$

and the result follows by Theorem 4.3. ∎

Lemma 4.20 shows that the problem of finding a stabilizing controller which minimizes the combination of (unweighted) objectives S, KS, SG and KSG (where $S = (I - GK)^{-1}$) has an exact solution, and further, it follows that any controller solving the normalized LCF robust stabilization problem in Proposition 4.1, will also solve (4.42)

Remark 4.21

If we were now to apply the Glover and Doyle (1988) solution approach to the four objective problem in (4.42), the corresponding Riccati equations for X_∞ and Y_∞ are

$$\begin{aligned} X_\infty A + A^* X_\infty - \frac{\gamma^2 - 1}{\gamma^2} X_\infty BB^* X_\infty + \frac{\gamma^2}{\gamma^2 - 1} C^*C &= 0 \\ Y_\infty A^* + AY_\infty - Y_\infty C^*CY_\infty + BB^* &= 0 \end{aligned}$$

and it is simple to verify that

$$\begin{aligned} X_\infty &= \frac{\gamma^2}{\gamma^2 - 1} X \\ Y_\infty &= Z \end{aligned}$$

are the stabilizing solutions. Note that $X_\infty \geq 0$ and $Y_\infty \geq 0$ for all $\gamma > 1$, and by Theorem 3.14(a) we have that a solution exists if

$$\rho(X_\infty Y_\infty) \leq \gamma^2$$

$$\Rightarrow \gamma^2 \geq 1 + \lambda_{max}(XZ),$$

which is the result of (4.29) again.

Note from Table 3.1 in Chapter 3 that transfer functions S, SG are typically used as performance objectives, while KS, KSG reflect robust stability properties. As discussed in Section 3.4, a more effective design problem is one in which these transfer function objectives are frequency weighted to trade-off between performance and robust stability over different frequency ranges. The main result of this section shows that for a special class of weighting selections, we can obtain an exact solution to a weighted version of the problem in Lemma 4.20.

First, we will generalize the normalized LCF robust stabilization problem by augmenting the nominal plant, G, with invertible (in R) pre and post compensators W_1 and W_2 respectively, such that an augmented plant G_A is given by

$$G_A = W_2 G W_1$$

and we assume W_1, W_2 are such that G_A has no hidden unstable modes.

The new design objective is to find a stabilizing controller, K_A, which robustly stabilizes the normalized LCF of G_A given by $(\tilde{N}_A, \tilde{M}_A)$ where

$$G_A = \tilde{M}_A^{-1} \tilde{N}_A$$

(See Figure 4.3a). We now give the weighted version of Lemma 4.20:

Theorem 4.22

$$\inf_{K \text{ stabilizing}} \left\| \begin{bmatrix} W_1^{-1} K \\ W_2 \end{bmatrix} (I - GK)^{-1} [W_2^{-1}, G W_1] \right\|_\infty = \gamma_A \tag{4.43}$$

and

$$\gamma_A \triangleq (1 - \|[\tilde{N}_A, \tilde{M}_A]\|_H^2)^{-1/2} \tag{4.44}$$

where $(\tilde{N}_A, \tilde{M}_A)$ is a normalized left coprime factorization of $G_A = W_2 G W_1$.

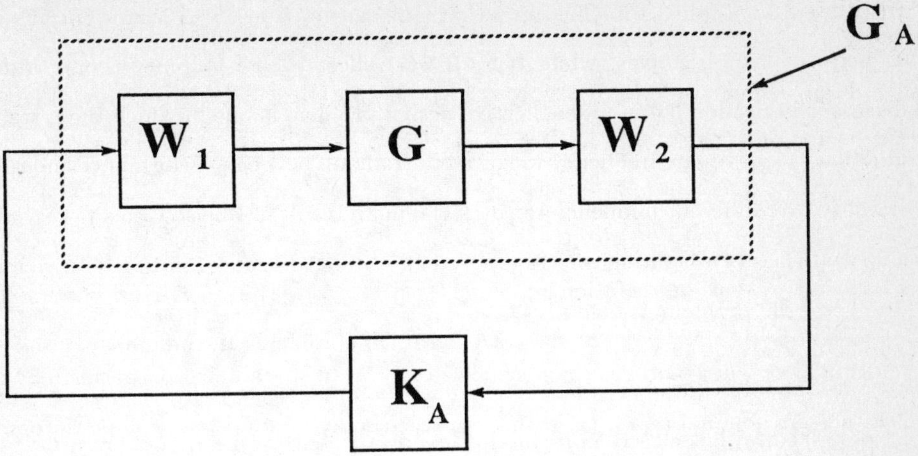

(a) Shaped Plant

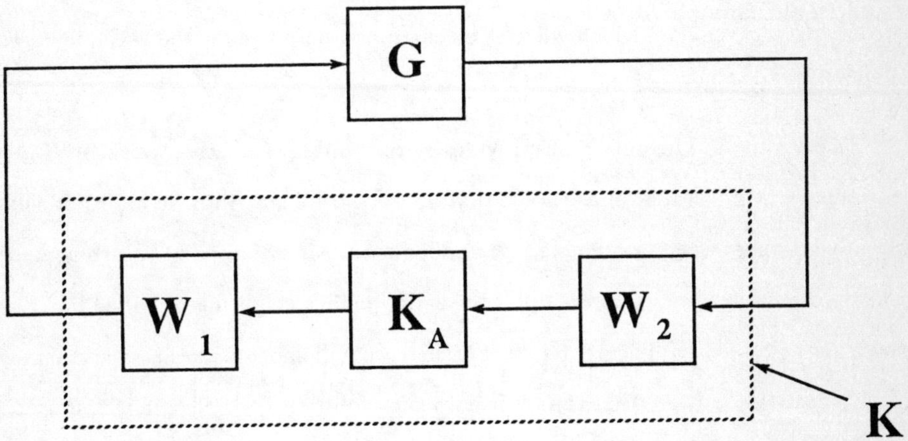

(b) Final Controller

Figure 4.3 Augmented Problem

Proof:

From Lemma 4.20, and Figure 4.3(a), the normalized LCF robust stabilization problem applied to G_A can be written as: *find*

$$\inf_{K_A \ stabilizing} \left\| \begin{bmatrix} I \\ K_A \end{bmatrix} (I - G_A K_A)^{-1} [I, G_A] \right\|_\infty ,$$

which has an optimal solution $\gamma_A \triangleq (1 - \|[\tilde{N}_A, \tilde{M}_A]\|_H^2)^{-1/2}$. Now, noting from Figure 4.3(b) that $K \triangleq W_1 K_A W_2$ is a controller stabilizing the nominal plant, simple substitution for G_A and K_A, and some minor manipulations, yields the required result in (4.43). Note that closed-loop stability is not affected by the change from $G_A \rightarrow G$ and $K_A \rightarrow K$. ■

Theorem 4.22 shows that we have the freedom to select weighting functions W_1 and W_2 for this four objective problem, and for any selection, an exact solution will exist for the minimum achievable norm in (4.44). (If a less restricted weighting arrangement is desired, then the iterative approach of Section 3.4 will need to be used.) An interesting problem, which will not be examined here, is how the variable γ_A varies with weighting function selection.

The result in Theorem 4.22 provides strong links with other work in H_∞ optimization. Grimble, (1988) has shown that an H_∞ problem will have an exact solution provided the cost weights and noise spectrum matrices in a related LQG problem have a particular relationship. This is similar to the frequency weighting provided by W_1 and W_2 in the problem of Theorem 4.22. Further, we will show next that a special choice of W_1 and W_2 can be used to obtain an exact solution to the so called 'Parametric Mixed Sensitivity Problem' studied by Kasenally and Limebeer, (1988).

The Parametric Mixed Sensitivity Problem, as defined in Kasenally and Limebeer, (1988) is the following: '*find*

$$\inf_K \left\| \begin{bmatrix} (1-\epsilon)GK \\ \epsilon I \end{bmatrix} (I - GK)^{-1} \right\|_\infty$$

where K is chosen from all stabilizing controllers.' In this problem, no dynamic weighting is used, but the parameter ϵ is used to trade off the relative importance of the two closed-loop transfer functions. We assume here that the plant G is invertible.

The solution to this problem is the following:

Lemma 4.23

The Parametric Mixed Sensitivity Problem has the following solution:

$$\inf_{K} \left\| \begin{bmatrix} (1-\epsilon)GK \\ \epsilon I \end{bmatrix} (I-GK)^{-1} \right\|_{\infty} = \frac{\epsilon(1-\epsilon)}{\mu}\gamma_A \tag{4.45}$$

where

$$\gamma_A \triangleq (1 - \|[\tilde{N}_A, \tilde{M}_A]\|_H^2)^{-1/2},$$

$$\mu \triangleq (2\epsilon^2 - 2\epsilon + 1)^{1/2}$$

and $(\tilde{N}_A, \tilde{M}_A)$ *is a normalized LCF of* GW_1 *where* W_1 *is a stable, minimum phase function satisfying*

$$GW_1W_1^*G^* = (1-\epsilon)^{-2}I.$$

Proof:

First, note that the problem in (4.43) can be rewritten, without loss of generality, as

$$\inf_{K \text{ stabilizing}} \left\| \begin{bmatrix} W_1^{-1}K \\ W_2 \end{bmatrix} (I-GK)^{-1}(W_2^{-1}W_2^{*-1} + GW_1W_1^*G^*)^{1/2} \right\|_{\infty} = \gamma_A \tag{4.46}$$

Now, select W_1 so that it is a minimum phase, stable function solving

$$GW_1W_1^*G^* = (1-\epsilon)^{-2}I.$$

That is, W_1^{-1} is a spectral factor of G scaled by a factor of $(1-\epsilon)$. Also, select $W_2 = \epsilon I$, and then in (4.46) we have that

$$\inf_{K \text{ stabilizing}} \left\| \begin{bmatrix} (1-\epsilon)GK \\ \epsilon I \end{bmatrix} (I-GK)^{-1}(\epsilon^{-2}I + (1-\epsilon)^{-2}I)^{1/2} \right\|_{\infty} = \gamma_A \tag{4.47}$$

where $\gamma_A = (1 - \|[\tilde{N}_A, \tilde{M}_A]\|_H^2)^{-1/2}$, and $(\tilde{N}_A, \tilde{M}_A)$ is a normalized LCF of GW_1. Note that as K stabilizes G, and as W_1, $W_1^{-1} \in RH_\infty$, closed-loop stability is not affected by replacing W_1^{-1} by G in (4.47). Finally, scaling both sides in (4.47) by

$$(\epsilon^{-2}I + (1-\epsilon)^{-2}I)^{-1/2} = \epsilon(1-\epsilon)/\mu,$$

the result in (4.45) is given. ∎

Remark 4.24

The result in Lemma 4.23 can in fact be generalized to the following weighted Mixed Sensitivity problem: find

$$\inf_{K \text{ stabilizing}} \left\| \begin{bmatrix} W_T GK \\ W_S \end{bmatrix} (I - GK)^{-1} (W_S^{-1} W_S^{*-1} + W_T^{-1} W_T^{*-1})^{1/2} \right\|_\infty$$

where W_T and W_S are any invertible, stable, minimum phase transfer functions. We can show that this problem has an exact solution in a similar way to the proof of Lemma 4.23, by requiring that W_1 and W_2 in (4.46) satisfy $GW_1 W_1{}^* G^* = W_T^{-1} W_T^{*-1}$ and $W_2 = W_S$.

4.5.2 Robust Stability of Passive Systems

In this final section we will show that if the nominal plant admits a *passive* state-space realization (A, B, C, D) (or equivalently if its transfer function G is *positive real*) then particularly attractive robust stabilization properties hold for the normalized left coprime factorization design problem.

Firstly, passivity is defined.

Definition 4.25 (Willems, 1972)

Let G be a stable transfer function of dimension $p \times p$ with a minimal state-space realization (A, B, C, D). This realization is passive if and only if it satisfies the following Linear Matrix Inequality:

$$\begin{bmatrix} -(A^*Q+QA) & C^*-QB \\ C-B^*Q & D+D^* \end{bmatrix} \geq 0$$

for some $Q = Q^* > 0$

or equivalently

(A, B, C, D) *is passive if and only if* $G = C(sI - A)^{-1}B + D$ *is positive real. i.e.*

(i) $G(s)$ *is real for all* $s \in R$

(ii) $G(s) + G^*(-\bar{s}) \geq 0$

for all $Re(s) \geq 0$, *and* s *not a pole of* G.

We will now state a result originally due to Opdenacker and Jonckheere, (1988), and extended here to the case $D \neq 0$, which relates passivity to the unique, positive definite solutions of the GCARE and GFARE of a minimal system (A, B, C, D).

Theorem 4.26

Let (A, B, C, D) *be a passive, degree* n, *state-space realization and define* $\mu_i \triangleq \lambda_i^{1/2}(ZX)$ *where* X, *(resp.* Z*) is the unique, positive definite solution to GCARE (resp. GFARE). Then*

$$\mu_i \leq 1 \quad i = 1, \ldots, n.$$

(See Appendix C for proof.)

The following Corollary is an immediate consequence of Theorem 4.26 and shows that passivity is also related to the singular values, σ_i, of the normalized left coprime factorization, $[\tilde{N}, \tilde{M}]$.

Corollary 4.27

If $G \stackrel{s}{=} (A, B, C, D)$ *is a passive, degree* n, *state-space realization, then*

$$\sigma_i \leq \frac{1}{\sqrt{2}} \quad i = 1, \ldots, n.$$

Proof:

This is immediate from Theorem 4.26 and by noting from Lemma 4.13 that

$$\sigma_i^2 \triangleq \lambda_i\{ZX(I+ZX)^{-1}\} = \frac{\lambda_i(ZX)}{1+\lambda_i(ZX)}$$
$$= \frac{\mu_i^2}{1+\mu_i^2}.$$

■

Remark 4.28

The proof of Theorem 4.26 given in Appendix C is based on the state-space approach of Opdenacker and Jonckheere (1988); however a direct proof of Corollary 4.27 (and hence Theorem 4.26) is possible via Lemma 4.20 and using the explicit controller $\mathbf{K} = -\mathbf{I}$. Firstly the resulting system will be stable since G is stable and

$$|\det(\mathbf{I} + \alpha G(j\omega))|^2 = \det(\mathbf{I} + \alpha(G+G^*) + \alpha \mathbf{G}\mathbf{G}^*)$$
$$> 1 \quad \text{for all } \omega \in R, \alpha \in [0,1]$$

by Definition 4.25 (ii). Thus no poles can cross the imaginary axis as α increases from 0 to 1. Secondly

$$G + G^* \geq 0 \Rightarrow \mathbf{I} + \mathbf{G}\mathbf{G}^* \leq (\mathbf{I}+G)(\mathbf{I}+G^*)$$
$$\Rightarrow (\mathbf{I}+G)^{-1}(\mathbf{I}+\mathbf{G}\mathbf{G}^*)(\mathbf{I}+G^*)^{-1} \leq \mathbf{I}$$
$$\Rightarrow \left\| \begin{bmatrix} \mathbf{I} \\ -\mathbf{I} \end{bmatrix} (\mathbf{I}+G)^{-1} [\mathbf{I} \quad \mathbf{G}] \right\|_\infty \leq \sqrt{2}$$

and the result follows from Lemma 4.20.

Positive real systems are known to be very easy to control and this result verifies that this is also true with respect to our present criterion, giving $\epsilon_{max} > \frac{1}{\sqrt{2}}$.

4.6 Review of Results

In this chapter we have given a detailed analysis of the normalized LCF robust stabilization problem. The solution of this problem has been shown to be directly linked to the solution of a related Nehari extension problem, and it has been shown that the optimal stability margin for this problem can be evaluated exactly without iterative methods. The general H_∞ solution approach of Glover and Doyle, (1988) confirms this exact result. In Sections 4.5 it was shown that the framework of the normalized LCF robust stabilization problem can be exploited to solve a class of H_∞ optimization problems. The special case where a nominal plant admits a passive state-space realization was also examined. In the next chapter we consider the problem of model reduction in this normalized coprime factor framework.

CHAPTER FIVE

REDUCED ORDER CONTROLLER DESIGN

5.1 Introduction

This chapter is concerned with the design of reduced order controllers via order reduction of a coprime factor representation of a given plant or controller. This approach was first proposed in Liu and Anderson, (1986), and has also been examined in Anderson and Liu, (1987), and Meyer (1987), (1988).

A major difficulty in reducing the controller degree is to ensure that the reduction error does not adversely affect closed-loop objectives. With this problem in mind, this chapter considers the coprime factor model reduction approach in conjunction with the controller design method introduced in Chapter 4 where a controller is selected to robustly stabilize the normalized coprime factor description of a nominal plant. In this case it turns out that a small model reduction error has a predictably small effect on the closed-loop objectives, and hence the reduction procedure and control design are in some sense compatible.

We will consider two different reduced order controller design approaches here; both of which fit neatly into the coprime factor framework introduced in Chapter 4:

(i) Model reduction of the normalized coprime factor representation of a nomimal plant, and then the design of a reduced order controller to robustly stabilize the normalized coprime factors of this reduced order plant.

(ii) Design of a controller to robustly stabilize normalized coprime factors of a nominal plant, and subsequent model reduction of the coprime factors of this controller giving a coprime factorization of a reduced order controller.

Hence, the essential difference between the two approaches in (i) and (ii) is the order in which the model reduction and robust stabilization design occur. In this chapter we will consider *optimal* controllers for the normalized LCF robust stabilization problem , as the results more clearly demonstrate the links between robust stabilization and model reduction. However, in Remark 5.12 we will discuss the use of sub-optimal controllers in this context.

The chapter is arranged in the following way. In Section 5.2 we give a brief overview of two common model reduction techniques, Balanced Truncation reduction and Hankel Norm Aproximation, and then review previous work on model reduction of coprime factors. In Section 5.3 we examine the two reduced order controller design methods described in (i) and (ii), and in each case, an explicit relationship can be derived relating the error involved in model reduction and the stability of the resulting closed-loop. Conclusions are given in Section 5.4.

5.2 Review

5.2.1 Model Reduction Techniques

A brief review of two common model reduction techniques, Balanced Truncation Approximation (BTA) and Hankel Norm Approximation (HNA), will now be given, and only those results central to this chapter are stated. The reader is referred to Moore, (1981) and Pernebo and Silverman, (1982) for full details on the Balanced Truncation method, and to Glover, (1984) for a complete study of the Hankel Norm Approximation problem.

We will firstly outline what a balanced realization is. A balanced realization is defined in the following way:

Definition 5.1 (Moore, 1981)

Let (A, B, C, D) be a minimal realization of an asymptotically stable transfer function G, then (A, B, C, D) is called **balanced** *if for*

$$P = \int_0^\infty e^{tA} BB^* e^{tA^*}\, dt$$
$$Q = \int_0^\infty e^{tA^*} C^* C e^{tA} dt$$

we have $P = Q = \Sigma \triangleq \mathrm{diag}(\sigma_1, \sigma_2, \ldots, \sigma_n)$. Σ *is called the Gramian of the system* (A, B, C, D). *The positive numbers* $\sigma_1, \sigma_2, \ldots, \sigma_n$ *are called the singular values of the system* (A, B, C, D) *where,* $\sigma_1 \geq \sigma_2 \ldots \geq \sigma_n > 0$.

In a balanced realization, the values of σ_i for $i = 1, \ldots, n$ have a special interpretation: each σ_i is associated with a particular state x_i of the balanced state-space system, and the relative size of σ_i indicates to what extent x_i contributes to the input/output behaviour of the system. For example, if $\sigma_k \gg \sigma_{k+1}$ then x_{k+1} will affect the input/output behaviour much less than x_k. This property is exploited in both Balanced Truncation and Hankel Norm Approximation where those states having little effect on the input/output behaviour of a system are discarded.

The basic idea of order reduction using balanced realizations is to consider an asymptotically stable, balanced, degree n system (A, B, C, D) and to partition it conformally as

$$A = \begin{bmatrix} A_{11} & A_{12} \\ A_{21} & A_{22} \end{bmatrix}, \quad B = \begin{bmatrix} B_1 \\ B_2 \end{bmatrix}, \quad C = [C_1, C_2]$$

such that for $1 \leq k < n$, $A_{11} \in \Re^{k \times k}$, $B_1 \in \Re^{k \times m}$, $C_1 \in \Re^{p \times k}$. The subsystem $G_r \stackrel{s}{=} (A_{11}, B_1, C_1, D)$ is considered as an approximant of (A, B, C, D), and it was shown in Pernebo and Silverman (1982) that this scheme has the important property that G_r is minimal and asymptotically stable if $\sigma_k > \sigma_{k+1}$.

Theoretically, Hankel Norm Approximation is slightly more involved, and the construction for a reduced order approximant of (A, B, C, D) is given in Glover, (1984, Theorem 7.2). The reduced order approximation of G, denoted $G_r \stackrel{s}{=} (A_r, B_r, C_r, D_r)$, is chosen so that the Hankel norm (see Definition 2.2) of the error system $(G - G_r)$ is minimized. The D_r term, which is independent of this minimization, can be chosen to satisfy additional constraints. An advantage of this approach over the Balanced Truncation approach is that the known H_∞ bound on the approximation error is smaller. For each of these approaches a bound on the H_∞ norm of the error bound can be obtained in terms of the Hankel singular values of G:

Lemma 5.2

Let G be of degree n, with associated (Hankel) singular values $\sigma_1 \leq \sigma_2 \ldots \leq \sigma_k < \sigma_{k+1} \ldots \leq \sigma_n < 0$, and let G_r be a degree k approximation of G. Then,

(a)(Glover, 1984; Enns, 1984): If G_r is obtained by Balanced Truncation,

$$\|G - G_r\|_\infty \leq 2 \sum_{i=k+1}^{n} \sigma_i.$$

(b) (Glover, 1984): If G_r is obtained by the Hankel Norm Approximation scheme in Glover, (1984, Theorem 7.2), then there exists a D_r such that

$$\|G - G_r\|_\infty \leq \sum_{i=k+1}^{n} \sigma_i.$$

These bounds enable model reduction effects to be quantified, and as we will show in Section 5.3, allows them to be incorporated into robust stability analysis.

It is important to note that both of these techniques require that G be *asymptotically stable*, which is often a restrictive condition.

A final note to make here is that frequency weighting can be incorporated into such model reduction schemes to emphasise frequency ranges over which it is desired to preserve the nominal input/output properties as much as possible during approximation. This problem has been studied for example by Enns, (1984), Latham and Anderson, (1985), Glover, (1986b) and recently by Al Saggaf and Franklin, (1988).

5.2.2 Coprime Factor Model Reduction

Model reduction in a coprime factor framework can be described as follows: (We give the Left Coprime Factor case here, and it should be noted that the dual Right Coprime Factor case follows identically.)

Procedure 5.3 *'The Coprime Factor Model Reduction Procedure'*

(1) Write G, the transfer function to be reduced (with degree n), as $G = \tilde{M}^{-1}\tilde{N}$ where $\tilde{M}, \tilde{N} \in RH_\infty$ are left coprime factors of G.

(2) Using an appropriate model reduction technique, approximate $[\tilde{N}, \tilde{M}]$ of degree n, by $[\tilde{N}_r, \tilde{M}_r]$ of degree k, $k < n$.

(3) Form the reduced order transfer function G_r (with degree k) by $G_r = \tilde{M}_r^{-1}\tilde{N}_r$.

Noting that several model reduction techniques such as Balanced Truncation, Hankel Norm Approximation require that the model to be approximated be stable, an immediate appeal of this approach is that, by definition, the coprime factors of *any* minimal system are always asymptotically stable. This means that we can reduce the order of both stable and unstable systems in this framework. Also note that G and G_r can have different numbers of RHP (right half plane) poles, as this just corresponds to $\tilde{M}$ and $\tilde{M}_r$ having differing numbers of RHP zeros.

We now briefly review the work in this area to date.

Liu and Anderson, (1986), consider model reduction of LQG-type controllers, where a right coprime factorization (RCF) of the controller is formed using state feedback and observer gains following the approach of Nett *et al* (1984). Given (U, V), the RCF of the controller, K, an unweighted Balanced Truncation approximation of the function $\begin{bmatrix} U \\ V \end{bmatrix}$ is performed, and following Procedure 5.3, a reduced order controller is formed. A condition is given which determines whether a reduced order controller of a given degree is guaranteed to satisfy closed-loop stability constraints. Also, a measure for robust performance is given in terms of a bound on the H_∞ norm of the change in the closed-loop transfer function $\mathcal{H}(G, K)$ denoted

$$\mathcal{H}(G, K) \triangleq \begin{bmatrix} I \\ K \end{bmatrix}(I - GK)^{-1}[G, I] + \begin{bmatrix} 0 & -I \\ 0 & 0 \end{bmatrix}, \tag{5.1}$$

when the nominal controller is replaced by a reduced order one. In a further paper by the same authors, Anderson and Liu, (1987), this work is extended beyond an LQG framework, and a *weighted* coprime factor reduction technique is suggested which allows the reduced order controller to be obtained while still guaranteeing closed-loop stability.

An advantage of using Balanced Truncation Approximation in coprime factor model reduction is that, by Lemma 5.2, it is possible to obtain an upper bound on the truncation error. This result is a useful indication of how many states can be removed without extensively changing the input/output properties of $[\tilde{N}, \tilde{M}]$.

Meyer (1987), (1988), considers unweighted Balanced Truncation of the *normalized* coprime factors of the nominal plant. A particular advantage of this approach is the following result:

Lemma 5.4 (Meyer, 1987)

Given $(\tilde{N}, \tilde{M})$, a normalized LCF of G, of degree n, let $[\tilde{N}_r, \tilde{M}_r]$ be a degree k $(k < n)$ Balanced Truncation approximation of $[\tilde{N}, \tilde{M}]$, which has Hankel singular values $\sigma_1 \geq \sigma_2 \ldots \sigma_k > \sigma_{k+1}, \ldots \geq \sigma_n > 0$. Then $(\tilde{N}_r, \tilde{M}_r)$ is a normalized left coprime factorization of G_r where $G_r = \tilde{M}_r^{-1} \tilde{N}_r$, and $[\tilde{N}_r, \tilde{M}_r]$ has Hankel singular values given by $\sigma_1, \sigma_2 \ldots \sigma_k$.

In Section 5.3, we will show that this result is useful when designing reduced order controllers. Further, it is relatively easy to show (see Ober and McFarlane, 1988) that Meyer's plant reduction scheme is *identical* to the *closed-loop*, 'Riccati Balancing ' model reduction procedure of Jonckheere and Silverman, (1983). In Riccati Balanced model reduction, a realization of G is chosen so that the solutions to GCARE and GFARE are identical and satisfy $X = Z = diag(\mu_1, \ldots, \mu_n)$ and the states corresponding to small 'μ_i' are discarded to form G_r. Normalized coprime factor model reduction can be shown to preserve closed-loop properties, and this will be a central consideration in this chapter.

5.3 Reduced Order Controller Design

5.3.1 Problem Statement

The aim of closed-loop model reduction is to generate a reduced order controller which stabilizes a given nominal plant. In this section we will consider reduced order controller design in conjunction with the normalized LCF robust stabilization problem studied in Chapter 4. As mentioned earlier, there are two different approaches that can be taken in solving this problem, and these are now formally stated:

Procedure 5.5 ***'The Nominal Plant Model Reduction Procedure'***

Consider a nominal plant G*, of degree* n*, with associated normalized LCF given by* $(\tilde{N}, \tilde{M})$*, where* $[\tilde{N}, \tilde{M}]$ *has (Hankel) singular values* $1 > \sigma_1 \geq \sigma_2 \geq \ldots \geq \sigma_{k+r} > \sigma_{k+r+1} \cdots \geq \sigma_n > 0$ *and* r *is the multiplicity of* σ_1*. Let* $[\tilde{N}_r, \tilde{M}_r]$ *be a degree* $k+l$ $(k+l<n)$ *approximation of* $[\tilde{N}, \tilde{M}]$ *with* l *the multiplicity of* $\sigma_1([\tilde{N}_r, \tilde{M}_r])$*, and* $G_r \triangleq \tilde{M}_r^{-1}\tilde{N}_r$ *is a degree* $k+l$ *approximation of* G*. Then, a degree* k *reduced order controller for* G *is given by* K_{r_1}*, which optimally robustly stabilizes the normalized LCF of* G_r*.*

Procedure 5.6 ***'The Controller Model Reduction Procedure'***

Consider a degree $n-r$ *controller,* K*, which optimally robustly stabilizes the normalized LCF of the nominal plant* G *with stability margin* ϵ_{max}*. Let* (U, V) *be an RCF of* K *such that* $\begin{bmatrix} U \\ V \end{bmatrix}$ *is an optimal Nehari extension of* $\begin{bmatrix} -\tilde{N}^* \\ \tilde{M}^* \end{bmatrix}$ *with tolerance* $(1-\epsilon_{max}^2)^{1/2}$*, and let* $\begin{bmatrix} U_r \\ V_r \end{bmatrix}$ *be a degree* k $(k < n-r)$ *approximation of* $\begin{bmatrix} U \\ V \end{bmatrix}$ *which has (Hankel) singular values given by* $\sigma_{c_1} \geq \sigma_{c_2} \ldots \sigma_{c_k} > \sigma_{c_{k+1}} \cdots \geq \sigma_{c_{n-r}} > 0$*. Then* $K_{r_2} \triangleq U_r V_r^{-1}$ *is a degree* k *approximation of* K*, and a reduced order controller for* G*.*

Either of these two approaches can be used to generate reduced order controllers for a nominal plant, and in the next two sections we will establish conditions whereby such controllers are stabilizing. Note that in Procedures 5.5 and 5.6 we do not specify the particular technique used in performing the model reduction. This selection can be made by the designer, and we will show later that, for example, there are particular advantages in choosing Balanced Truncation Approximation. Further, although a reduced order controller is obviously no longer the optimal controller for the normalized coprime factor problem, it will be shown that it is still possible to guarantee a (reduced) stability margin. The problem of Procedure 5.5 is considered first:

5.3.2 Reduced Order Controllers by Plant Model Reduction

In this section we will consider reduced order controllers which are obtained by a normalized LCF robust stabilization design on a reduced order plant G_r as outlined in Procedure 5.5.

The following results refer to the stability of the closed-loop system incorporating the nominal plant G and reduced order controller K_{r_1}. Following Procedure 5.5, we denote $(\tilde{N}, \tilde{M})$ as the normalized LCF of G, the nominal plant, and $(\tilde{N}_r, \tilde{M}_r)$ as an LCF of G_r, where $[\tilde{N}_r, \tilde{M}_r]$ is a degree $k + l$ approximation of $[\tilde{N}, \tilde{M}]$. Define the model reduction error by

$$\epsilon_1 \triangleq \|[\tilde{N} - \tilde{N}_r, \tilde{M} - \tilde{M}_r]\|_\infty \tag{5.2}$$

and recall from Chapter 4 that the optimal controller K is chosen such that

$$\left\| \begin{bmatrix} K \\ I \end{bmatrix} (I - GK)^{-1} \tilde{M}^{-1} \right\|_\infty = \epsilon_{max}^{-1}$$

where $\epsilon_{max} = (1 - \sigma_1^2)^{1/2}$ and σ_1 is the largest Hankel singular value (or Hankel norm) of $[\tilde{N}, \tilde{M}]$.

Theorem 5.7

A degree k controller, K_{r_1}, obtained from the Nominal Plant Model Reduction Problem of Procedure 5.5, with ϵ_1 given by (5.2), stabilizes the nominal plant, G, provided

$$\epsilon_1 < \frac{\epsilon_{max}}{2 + \epsilon_{max}}$$

and the system has a reduced order stability margin, ϵ_{r_1}, satisfying

$$\epsilon_{r_1} \geq \frac{(1-\epsilon_1)}{(1+\epsilon_1)}(\epsilon_{max}-\epsilon_1)-\epsilon_1 .$$

Proof:

Consider a degree $n-r$ controller, K, which robustly stabilizes the normalized LCF of G, with stability margin ϵ_{max}. Then, let $[\tilde{N},\tilde{M}]$, of degree n, be approximated by $[\tilde{N}_r,\tilde{M}_r]$, of degree $k+l$, with associated reduction error, ϵ_1, and where $G_r=\tilde{M}_r^{-1}\tilde{N}_r$. Then, provided $\epsilon_1<\epsilon_{max}$, K robustly stabilizes the LCF $(\tilde{N}_r,\tilde{M}_r)$ with a stability margin of at least $(\epsilon_{max}-\epsilon_1)$. Using the Small Gain Theorem this implies that

$$\left\| \begin{bmatrix} K \\ I \end{bmatrix} (I-G_rK)^{-1}\tilde{M}_r^{-1} \right\|_\infty \leq (\epsilon_{max}-\epsilon_1)^{-1} \tag{5.3}$$

Next, noting that $(\tilde{N}_r,\tilde{M}_r)$ may not be a *normalized* LCF of G_r we require a condition similar to (5.3) for the normalized LCF of G_r. We now obtain the corresponding normalized LCF of G_r:

First, define a spectral factor W: $W,\ W^{-1}\in RH_\infty$, which solves

$$W^{-1}W^{*-1}=\tilde{M}_r\tilde{M}_r^{\ *}+\tilde{N}_r\tilde{N}_r^{\ *}. \tag{5.4}$$

Then the normalized LCF of G_r denoted by $(\tilde{N}_{rn},\tilde{M}_{rn})$ is given by

$$(\tilde{N}_{rn},\tilde{M}_{rn}) \stackrel{\triangle}{=} (W\tilde{N}_r,W\tilde{M}_r).$$

Next, defining the perturbations induced by model reduction by

$$\Delta_N \stackrel{\triangle}{=} \tilde{N}_r-\tilde{N}$$

and

$$\Delta_M \stackrel{\triangle}{=} \tilde{M}_r-\tilde{M},$$

we can estimate an upper bound on the H_∞ norm of W and W^{-1}.

Note that by (5.4),

$$\begin{aligned} W^{-1}W^{*-1} &= (\tilde{M}+\Delta_M)(\tilde{M}+\Delta_M)^* + (\tilde{N}+\Delta_N)(\tilde{N}+\Delta_N)^* \\ &= I + [\Delta_N, \Delta_M]\begin{bmatrix} \tilde{N}^* \\ \tilde{M}^* \end{bmatrix} + [\tilde{N}, \tilde{M}]\begin{bmatrix} \Delta_N^* \\ \Delta_M^* \end{bmatrix} + [\Delta_N, \Delta_M]\begin{bmatrix} \Delta_N^* \\ \Delta_M^* \end{bmatrix}. \end{aligned}$$

Then noting that $\|[\tilde{N}, \tilde{M}]\|_\infty = 1$, and that $\|[\Delta_N, \Delta_M]\|_\infty = \epsilon_1$, we have the following bounds:

$$\begin{aligned} \|W^{-1}\|_\infty &\le (1+\epsilon_1) \\ \|W\|_\infty &\le \frac{1}{(1-\epsilon_1)}. \end{aligned}$$

It then follows that

$$\begin{aligned} \|[\tilde{N}_r, \tilde{M}_r]\|_\infty &\le \|W^{-1}\|_\infty \|[\tilde{N}_{rn}, \tilde{M}_{rn}]\|_\infty \\ &\le (1+\epsilon_1). \end{aligned} \tag{5.5}$$

Multiplying each side of (5.3) by $[\tilde{N}_r, \tilde{M}_r]$ and using (5.5) we have

$$\begin{aligned} &\left\| \begin{bmatrix} K \\ I \end{bmatrix} (I - G_r K)^{-1} [I, G_r] \right\|_\infty \le (1+\epsilon_1)(\epsilon_{max} - \epsilon_1)^{-1} \\ \Rightarrow &\left\| \begin{bmatrix} K \\ I \end{bmatrix} (I - G_r K)^{-1} \tilde{M}_{rn}^{-1} \right\|_\infty \le (1+\epsilon_1)(\epsilon_{max} - \epsilon_1)^{-1}. \end{aligned} \tag{5.6}$$

Therefore, using a Small Gain Theorem argument again, the optimal controller, K_{r_1}, which robustly stabilizes the normalized LCF of G_r, must have a stability margin of at least $\frac{(\epsilon_{max}-\epsilon_1)}{(1+\epsilon_1)}$.

Reversing the previous argument using K_{r_1}, we then have that

$$\begin{aligned} &\left\| \begin{bmatrix} K_{r_1} \\ I \end{bmatrix} (I - G_r K_{r_1})^{-1} \tilde{M}_{rn}^{-1} \right\|_\infty \le (1+\epsilon_1)(\epsilon_{max} - \epsilon_1)^{-1} \\ \Rightarrow &\left\| \begin{bmatrix} K_{r_1} \\ I \end{bmatrix} (I - G_r K_{r_1})^{-1} \tilde{M}_r^{-1} W^{-1} \right\|_\infty \le (1+\epsilon_1)(\epsilon_{max} - \epsilon_1)^{-1} \\ \Rightarrow &\left\| \begin{bmatrix} K_{r_1} \\ I \end{bmatrix} (I - G_r K_{r_1})^{-1} \tilde{M}_r^{-1} \right\|_\infty \le (1+\epsilon_1)(\epsilon_{max} - \epsilon_1)^{-1} \|W\|_\infty \\ &\qquad \le \frac{(1+\epsilon_1)}{(1-\epsilon_1)}(\epsilon_{max} - \epsilon_1)^{-1}. \end{aligned}$$

Hence K_{r_1} guarantees closed-loop stability for all perturbations on $(\tilde{N}_r, \tilde{M}_r)$ achieving H_∞ norm less than $\frac{(1-\epsilon_1)}{(1+\epsilon_1)}(\epsilon_{max} - \epsilon_1)$. Now, to consider the closed-loop system incorporating G and K_{r_1}, we incorporate the error, ϵ_1, due to model reduction on $[\tilde{N}, \tilde{M}]$. We then have that K_{r_1}, the controller which robustly stabilizes the normalized LCF of G_r will stabilize G with a minumum guaranteed stability margin, ϵ_{r_1}, given by

$$\epsilon_{r_1} \geq \frac{(1-\epsilon_1)}{(1+\epsilon_1)}(\epsilon_{max} - \epsilon_1) - \epsilon_1$$

where to ensure $\epsilon_{r_1} > 0$, and hence guarantee nominal stability, we require

$$\epsilon_1 < \frac{(1-\epsilon_1)}{(1+\epsilon_1)}(\epsilon_{max} - \epsilon_1)$$

$$\Rightarrow \epsilon_1 < \frac{\epsilon_{max}}{2+\epsilon_{max}}.$$

■

Remark 5.8 ***Balanced Truncation***

We can improve the results of Theorem 5.7 by specifically choosing to use Balanced Truncation Approximation in Procedure 5.5 to find $(\tilde{N}_r, \tilde{M}_r)$. (Note that this choice gives $r = l$ in Procedure 5.5.) By Lemma 5.4, this has the advantage that $(\tilde{N}_r, \tilde{M}_r)$ will be a *normalized* LCF of G_r. Also, note that as the maximum singular value of $[\tilde{N}_r, \tilde{M}_r]$ is *identical* to that of $[\tilde{N}, \tilde{M}]$, the maximum achievable stability margin (ϵ_{max}) for G_r, will be the same as that for G. Hence the stability margin for the reduced order controller, K_{r_1}, with G is given by

$$\epsilon_{r_1} \geq \epsilon_{max} - \epsilon_1 \tag{5.7}$$

where we require that $\epsilon_1 < \epsilon_{max}$ to ensure closed-loop stability. Further, using the error bound given in Lemma 5.2, and recalling that $\epsilon_{max} = (1-\sigma_1^2)^{1/2}$, we have, by (5.7), that

$$\epsilon_{r_1} \geq (1-\sigma_1^2)^{1/2} - 2\sum_{i=k+r+1}^{n} \sigma_i$$

and closed-loop stability is guaranteed if

$$\sum_{i=k+r+1}^{n} \sigma_i < 0.5(1-\sigma_1^2)^{1/2}, \tag{5.8}$$

a condition which is given *entirely* in terms of the Hankel singular values of $[\tilde{N}, \tilde{M}]$. These results indicate that Balanced Truncation has particular advantages for this problem.

The next section examines the reduced order design problem of Procedure 5.6.

5.3.3 Reduced Order Controllers by Controller Model Reduction

We now consider the model reduction problem of Procedure 5.6. We examine the effects on the closed-loop stability and performance of replacing a full order optimal controller (designed to robustly stabilize the normalized LCF of G) by a reduced order approximation, K_{r_2}.

Using the notation given in Procedure 5.6, define the controller reduction error by

$$\epsilon_2 = \left\| \begin{bmatrix} U - U_r \\ V - V_r \end{bmatrix} \right\|_\infty$$

We now derive a corresponding result to that of Theorem 5.7 for the reduced order stability margin using this procedure.

Theorem 5.9

A degree k controller, K_{r_2}, obtained from the Controller Model Reduction Problem of Procedure 5.6, robustly stabilizes the nominal plant G provided

$$\epsilon_2 < 1 - (1 - \epsilon_{max}^2)^{1/2}$$

and the system has a reduced order stability margin, ϵ_{r_2}, given by

$$\epsilon_{r_2} \geq \{1 - ((1 - \epsilon_{max}^2)^{1/2} + \epsilon_2)^2\}^{1/2}. \tag{5.9}$$

Proof:

Consider an optimal controller, of degree $n - r$, for the normalized LCF robust stabilization problem. Then by Theorem 4.3, it has a RCF (U, V) satisfying

$$\left\| \begin{bmatrix} -\tilde{N}^* \\ \tilde{M}^* \end{bmatrix} + \begin{bmatrix} U \\ V \end{bmatrix} \right\|_\infty = (1 - \epsilon_{max}^2)^{1/2}.$$

Now consider K_{r_2}, a degree k approximation of K from Procedure 5.6, with RCF (U_r, V_r). Denoting ϵ_{r_2} to be the reduced stability margin for K_{r_2} with G, then by

Theorem 4.2 it can be shown that

$$\left\| \begin{bmatrix} K_{r_2} \\ I \end{bmatrix} (I - GK_{r_2})^{-1} \tilde{M}^{-1} \right\|_\infty = \epsilon_{r_2}^{-1}$$

$$\Leftrightarrow \left\| \begin{bmatrix} -\tilde{N}^* \\ \tilde{M}^* \end{bmatrix} + \begin{bmatrix} U_r \\ V_r \end{bmatrix} \right\|_\infty = (1 - \epsilon_{r_2}^2)^{1/2}.$$

Hence, using the triangle inequality,

$$\begin{aligned} (1 - \epsilon_{r_2}^2)^{1/2} &= \left\| \begin{bmatrix} -\tilde{N}^* \\ \tilde{M}^* \end{bmatrix} + \begin{bmatrix} U_r \\ V_r \end{bmatrix} \right\|_\infty \\ &\leq \left\| \begin{bmatrix} -\tilde{N}^* \\ \tilde{M}^* \end{bmatrix} + \begin{bmatrix} U \\ V \end{bmatrix} \right\|_\infty + \left\| \begin{bmatrix} U - U_r \\ V - V_r \end{bmatrix} \right\|_\infty \\ &\leq (1 - \epsilon_{max}^2)^{1/2} + \epsilon_2 \qquad (5.10) \\ \Rightarrow \epsilon_{r_2} &\geq \{1 - ((1 - \epsilon_{max}^2)^{1/2} + \epsilon_2)^2\}^{1/2} \end{aligned}$$

as required. Further, we note that

$$\epsilon_2 < 1 - (1 - \epsilon_{max}^2)^{1/2}$$

is a sufficient condition to guarantee closed-loop stability. ■

Remark 5.10 *Balanced Truncation*

We can also simplify the results in Theorem 5.9 for the specific case of Balanced Truncation in Procedure 5.6. Firstly recall that the stability margin associated with the full order controller is given by $\epsilon_{max} = (1 - \sigma_1^2)^{1/2}$. Next, by Theorem 4.3 and from a result that is due to Glover[1], (1984), we have

$$\sum_{i=k+1}^{n-r} \sigma_{c_i} \leq \sum_{i=k+r+1}^{n} \sigma_i$$

where σ_{c_i}, $i = k+1, \ldots, n-r$ denotes the singular values of the RCF of the full order optimal controller which has been truncated by model reduction. Then from (5.8) the reduced stability margin becomes

$$\epsilon_{r_2} \geq \{1 - (\sigma_1 + 2 \sum_{i=k+r+1}^{n} \sigma_i)^2\}^{1/2}$$

1 Glover, (1984, Theorem 6.3), shows that if $Q \in RH_\infty$ of degree $n - r$ is the optimal Nehari extension of R of degree n, where $R^* \in RH_\infty$, then by construction $\sigma_i(Q) \leq \sigma_{r+i}(R^*)$ $i = 1, \ldots, n - r$.

where the condition

$$2 \sum_{i=k+r+1}^{n} \sigma_i < 1 - \sigma_1 \tag{5.11}$$

guarantees closed-loop stability. Note that similar results in this approach are true when Hankel Norm approximation (see Glover, 1984) is adopted except that the factor of 2 may be removed.

Remark 5.11

The measure $\|\mathcal{H}(G,K)\|_\infty$, where $\mathcal{H}(G,K)$ is defined in (5.1), is a general indicator of closed-loop performance. By noting that

$$\|\mathcal{H}(G,K)\|_\infty \leq 1 + \left\| \begin{bmatrix} I \\ K \end{bmatrix} (I - GK)^{-1}[G, I] \right\|_\infty ,$$

it is simple to use the results of Theorems 5.7 and 5.9 to assess closed-loop performance with the reduced order controllers obtained by Procedures 5.5 and 5.6 respectively.

Remark 5.12 ***Sub-Optimal Reduced Order Controllers***

It is possible to obtain similar results to those given in Section 5.3.2 and in this section for the case when the robustly stabilizing controllers in Procedures 5.5 and 5.6 are sub-optimal controllers for the normalized left coprime factorization robust stabilization problem. An additional complication in such calculations is that the selection of the *sub-optimal stability margin* is constrained by the model reduction error. For example, consider the reduced order stability margin for Procedure 5.5 stated in Theorem 5.7. In the proof of this theorem we establish that a reduced order *optimal* controller K_{r_1} will robustly stabilize the reduced order plant, G_r, with stability margin $\frac{(\epsilon_{max}-\epsilon_1)}{(1+\epsilon_1)}$. Hence, a corresponding *sub-optimal* controller will robustly stabilize G_r with a guaranteed stability margin, $\frac{(\epsilon_{max}-\epsilon_1)}{(1+\epsilon_1)} - \Delta_\epsilon$, where Δ_ϵ indicates the reduction in stability margin due to the sub-optimal selection. We then require $\Delta_\epsilon < \frac{(\epsilon_{max}-\epsilon_1)}{(1+\epsilon_1)}$ to guarantee closed-loop stability of the reduced order closed-loop. Following the proof of Theorem 5.7, the guaranteed stability margin for the reduced order controller with the nominal plant becomes

$$\epsilon_{r_1} \geq \frac{(1-\epsilon_1)}{(1+\epsilon_1)} \left[(\epsilon_{max} - \epsilon_1) - \Delta_\epsilon(1+\epsilon_1)\right]^{-1} - \epsilon_1 ,$$

and it is possible to revise the result of Theorem 5.9 in a similar manner.

In practice however, when a sub-optimal controller is used in reduced order controller design, it is typical to select the associated stability margin ϵ to be such that $0.95\epsilon_{max} \leq \epsilon < \epsilon_{max}$, and under these circumstances, the results in Theorems 5.7 and 5.9, for optimal controllers, provide satisfactory information about reduced order stability margins.

Remark 5.13 ***Model Reduction of Passive Systems***

In Section 4.5.2 we introduced plants with passive state-space systems in the normalized LCF framework, and it can be shown that such systems are particularly good candidates for the reduced order controller design procedures that we have discussed in this chapter: For example, by Theorem 5.7, a reduced order controller K_{r_1} designed by the problem of Procedure 5.5 will always stabilize a positive real plant (that is, one with a passive realization) provided

$$\epsilon_1 \leq \frac{1}{2\sqrt{2}+1} = 0.2612.$$

If we then choose to use Balanced Truncation approximation for this problem we have that closed-loop stability is guaranteed if

$$\sum_{i=k+r+1}^{n} \sigma_i \leq \frac{1}{2\sqrt{2}} = 0.3536.$$

Similar results can be obtained for controllers designed by Procedure 5.6 using Theorem 5.8. Further, noting from Lemma 4.5 that $\sigma_i < 1$ for all $i = 1, \ldots, n$, the above results indicate that significant scope for model reduction exists when the nominal plant admits a passive state-space realization.

5.4 Summary

The results of this chapter have shown how to systematically reduce the controller degree. They clearly indicate how the maximum stability margin $(1-\sigma_1^2)^{1/2}$ influences allowable model reduction error, and also show how $\{\sigma_i\}$ can be used to determine the required degree to achieve this accuracy of approximation. The Balanced Truncation approach is well suited although it is difficult to make definitive statements since the bounds of Theorems 5.7 and 5.9 and equations (5.8) and (5.11) are based on conservative error analysis.

CHAPTER SIX

A LOOP SHAPING DESIGN PROCEDURE

6.1 Introduction

The normalized LCF robust stabilization problem introduced in Chapter 4 is a particularly simple approach to robust controller design. The trade-off for this simplicity however, is that there is no scope for designer inputs and, in particular, performance objectives cannot be included.

In this chapter we will introduce a procedure which allows performance requirements to be specified within the normalized LCF framework, and allows the designer to trade-off between performance and robust stability objectives. There are in fact several ways of doing this: one way of incorporating performance objectives is to follow the approach outlined in Chapter 3, where it was shown that *both* robust stability and performance objectives can be incorporated into a single H_∞ optimization problem. (See Example 3.13.) We could therefore append the normalized LCF robust stabilization objectives with an additional (weighted) performance objective in the H_∞ optimization problem in (4.3). A major disadvantage of such an approach is that, as we have changed the H_∞ problem, its solution would no longer be exact, and the computational requirements in controller design would therefore be greater. Another objection to such a method is that there is no clear methodology for weighting selection in H_∞ design, and selection of an appropriate weight on the performance objective would typically involve a number of 'trial and error' experiments. (The appropriate selection of weights over frequency and between objectives is not straightforward, and tends to be developed for each specific example. See, for example, Postlethwaite et al, 1987, Safonov and Chiang, 1988, and Limebeer and Kasenally, 1986.) Further, it is necessary that the weights on the desired closed-loop objectives should adequately reflect the properties of the nominal plant, and this may not be simple to achieve.

Noting the above objections to the 'standard' H_∞ approach to incorporating performance objectives, we are therefore proposing an alternative, less regular, H_∞ design approach in this chapter. In this approach we use the well known 'loop shaping' principles to introduce performance/robustness trade-offs, while using the normalized LCF robust stabilization method as a means of guaranteeing closed-loop stability. This procedure is briefly outlined below:

(i) Using pre and/or post compensation, the nominal plant singular values are 'shaped', using a simple loop shaping method, to give a desired open-loop shape. A 'shaped' plant is formed by combining the nominal plant and these compensators.

(ii) A feedback controller which robustly stabilizes the normalized LCF of the shaped plant is synthesised.

(iii) The feedback controller and shaping compensators are combined to form the final controller.

An advantage of this design approach is that we can include performance considerations (via loop shaping), yet still obtain an exact solution for the normalized LCF problem. Also, note that the designer makes no input (that is, selects no weights) in the H_∞ optimization in stage (ii), but rather selects the shaping functions in stage (i). In this procedure, the normalized LCF robust stabilization H_∞ problem is being used as a 'tool' to guarantee closed-loop stability while preserving a specified loop-shape. As will be discussed in the next section, this approach is similar in philosophy to the Loop Transfer Recovery approach to LQG design. It will be demonstrated that the loop shaping procedure is a systematic and conceptually simple way to specify design objectives.

This chapter gives a full description and analysis of this design method, and it will be shown that performance/robustness trade-offs can be made in a simple manner. In Section 6.2 we briefly examine the classical approach to loop shaping, and assess its benefits and limitations. In Section 6.3 the design method outlined above is formally stated, and then in Sections 6.4 and 6.5 we give theoretical justification for adopting this loop shaping based approach. In Section 6.4 we show that, ϵ, the stability margin for the normalized LCF robust stabilization problem can be used as a design indicator for this procedure, and in Section 6.5 we show that it is possible to make guarantees on closed-loop behaviour using this method. Section 6.6 examines the role of the shaping functions in design, indicating in particular how a performance/robust stability trade-off is achieved. Also, several other important aspects of the design method are examined. Concluding comments are given in Section 6.7.

6.2 Loop Shaping Methods

In Chapter 3 we showed that many performance and robust stability objectives can be written as requirements on the maximum singular values of particular closed-loop transfer functions. (Refer to Table 3.1.) The principal idea of 'Loop Shaping' is that the magnitude[1] (or maximum singular values) of these *closed-loop* transfer functions can be directly determined (over appropriate frequency ranges) by the singular values of the corresponding *open-loop* transfer function. Hence, the design of a controller, K, achieving certain specifications for the closed-loop system in Figure 6.1, can be achieved by selecting a K which appropriately 'shapes' the singular values of the open-loop system in Figure 6.2. (The reader is referred to Doyle and Stein, 1981, for a comprehensive introduction to loop shaping methods.)

To illustrate the relationship between common closed-loop design objectives and open-loop singular value requirements, we will consider four common design objectives and show that each objective can be *approximated*, over a specified frequency range, by an open-loop singular value.

In the following analysis we will use the well known singular value identities:

1 When referring to MIMO systems, recall that the maximum singular value, denoted $\overline{\sigma}(\bullet)$, is the maximum signal gain (or magnitude) at a given frequency, and that the minimum singular value, denoted $\underline{\sigma}(\bullet)$, is the corresponding minimum signal gain.

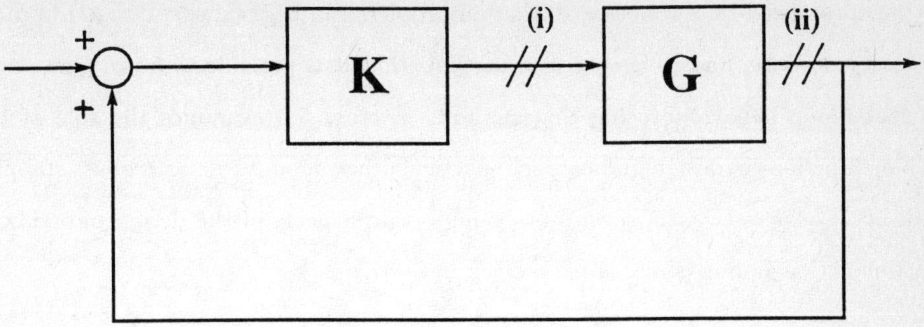

Figure 6.1 Closed-Loop System with Loop Break Points

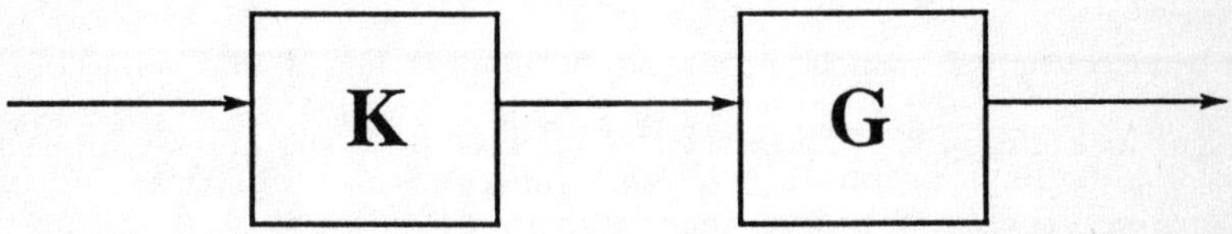

Figure 6.2 Open-Loop System

$\overline{\sigma}(A) = 1/\underline{\sigma}(A^{-1})$; $\overline{\sigma}(AB) \leq \overline{\sigma}(A)\overline{\sigma}(B)$; $\overline{\sigma}(A+B) \leq \overline{\sigma}(A) + \overline{\sigma}(B)$; $\underline{\sigma}(A+B) \geq \underline{\sigma}(A) - \overline{\sigma}(B)$. Also recall that the singular values are frequency dependent functions, and that the condition $\overline{\sigma}(A(j\omega)) \leq \|A\|_\infty$ is satisfied, by definition, for all ω.)

1) *Sensitivity*: recall from Table 3.1 that minimizing $\overline{\sigma}((I-GK)^{-1})$ minimizes the effect of output disturbances on the plant output. The following inequality relates this objective to an open-loop singular value condition:

$$\begin{aligned}\overline{\sigma}((I-GK)^{-1}) &= \frac{1}{\underline{\sigma}(I-GK)} \\ &\leq \frac{1}{\underline{\sigma}(GK)-1} \\ &\simeq \frac{1}{\underline{\sigma}(GK)} \text{ for frequencies: } \underline{\sigma}(GK) \gg 1. \end{aligned} \tag{6.1}$$

2) *Robustness in the face of additive uncertainty* on the nominal plant can be obtained by minimizing $\overline{\sigma}(K(I-GK)^{-1})$. We also have that:

$$\begin{aligned}\overline{\sigma}(K(I-GK)^{-1}) &\leq \frac{\overline{\sigma}(K)}{\underline{\sigma}(I-GK)} \\ &\simeq \overline{\sigma}(K) \text{ for frequencies: } \overline{\sigma}(GK) \ll 1. \end{aligned} \tag{6.2}$$

3) *Complementary Sensitivity*: minimizing $\overline{\sigma}(GK(I-GK)^{-1})$ minimizes the effect of sensor noise on the plant output, and also maximizes robust stability to multiplicative uncertainty at the plant output. Again, we have that:

$$\begin{aligned}\overline{\sigma}(GK(I-GK)^{-1}) &\leq \frac{1}{\underline{\sigma}((GK)^{-1})-1} \\ &\simeq \overline{\sigma}(GK) \text{ for frequencies: } \overline{\sigma}(GK) \ll 1. \end{aligned} \tag{6.3}$$

4) *Rejection of plant input disturbances at the plant output* will be maximized if $\overline{\sigma}((I-GK)^{-1}G)$ is minimized. Relating this to the open-loop system we have:

$$\begin{aligned}\overline{\sigma}((I-GK)^{-1}G) &= \overline{\sigma}\{((GK)^{-1}-I)^{-1}(GK)^{-1}G\} \\ &\simeq \overline{\sigma}(K^{-1}) \text{ if } K \text{ is square} \\ &= \frac{1}{\underline{\sigma}(K)} \text{ for frequencies: } \underline{\sigma}(GK) \gg 1. \end{aligned} \tag{6.4}$$

In each of the cases, 1)-4), we have approximated a closed-loop objective by a condition on the singular values of G and K over a particular frequency range. 1) and 4) are typically performance objectives, while 2) and 3) are robust stability objectives. Inspecting (6.1)-(6.4) it can be seen that the open-loop requirement for good performance ($\underline{\sigma}(GK)$, $\underline{\sigma}(K)$ large) is in conflict with the open-loop robust stability requirement ($\overline{\sigma}(GK)$, $\overline{\sigma}(K)$ small), demonstrating the well known design trade-off between performance and robust stability. Hence in terms of the measures in 1) - 4), this conflict indicates that we cannot expect to achieve arbitrarily good performance and robustness over all frequencies. However, we can achieve an acceptable compromise: as performance is typically most important at low frequency, we specify that $\underline{\sigma}(GK) \gg 1$ (i.e. $\underline{\sigma}(GK)$ large) for all $\omega \in (0, \omega_l)$, and as robust stability is typically most important at high frequency, we specify that $\overline{\sigma}(GK) \ll 1$ (i.e $\overline{\sigma}(GK)$ small) for all $\omega \in (\omega_u, \infty)$. The values ω_l and ω_u ($\omega_l < \omega_u$) are selected to indicate appropriate frequency ranges for these objectives.

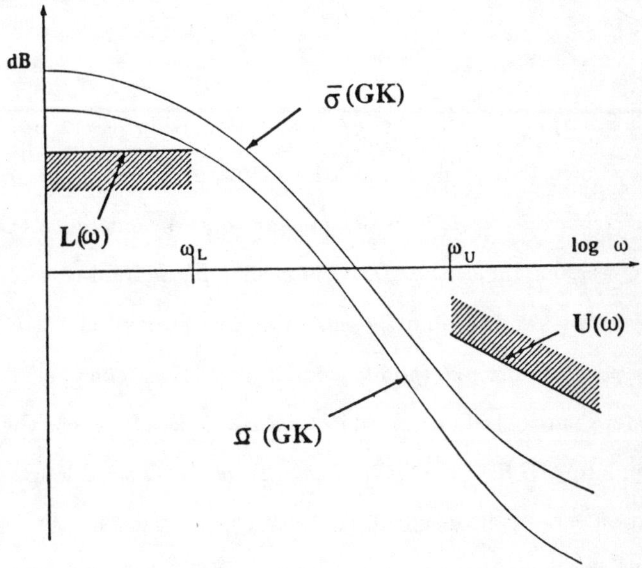

Figure 6.3 Open-Loop Singular Value Shaping

Figure 6.3 indicates graphically how the requirements on these closed-loop objectives constrain the shape of the open-loop singular values in design. The designer must choose K in such a way that $\overline{\sigma}(GK)$ and $\underline{\sigma}(GK)$ avoid the shaded regions. That is,

$\underline{\sigma}(GK) > L(\omega)$ for all $\omega \in (0, \omega_l)$ and $\overline{\sigma}(GK) < U(\omega)$ for all $\omega \in (\omega_u, \infty)$. (Additionally, it may be necessary to further adjust $\underline{\sigma}(K)$ over $(0, \omega_l)$, and $\overline{\sigma}(K)$ over (ω_u, ∞) to satisfy objectives 2) and 4) respectively.)

This singular value 'shaping' appears to be a particularly simple approach to design, involving only simple manipulations of the open-loop gain. However, shaping the open-loop system of Figure 6.2 does not guarantee that the closed-loop system of Figure 6.1 will be internally stable, as this condition cannot be determined from the open-loop singular values. For the SISO case, Bode, (1945), showed that, in general, closed-loop stability is dependent on the (open-loop) gain/phase relationship near crossover frequency ($|GK| = 1$), and that in particular, the rate of transition from high gain to low gain with frequency is limited by phase requirements. This means that, in Figure 6.3, the frequency range (ω_l, ω_u) cannot be chosen to be arbitrarily small, and it can be shown that for practical design, the maximum rate of transition (or *roll-off rate*) is -40db/decade. Loop shaping design is therefore complicated by this stability constraint.

In Doyle and Stein, (1981), this 'gain/phase relationship' is extended to MIMO systems, and it is shown that similar limitations exist in terms of the roll-off of the magnitude of the *eigenvalues* of GK (not *singular* values), denoted $\lambda_i(GK)$, in the crossover frequency region. Hence the MIMO loop shaping method is also complicated by stability requirements, and it is not possible to select parameters L, U, ω_l and ω_u in Figure 6.3 without considering stability constraints. This 'classical' loop shaping procedure is therefore particularly involved in the MIMO case, because the designer is required to manipulate $\underline{\sigma}(GK)$ and $\overline{\sigma}(GK)$ to achieve the desired loop shape, while restricting the roll-off rate of the eigenvalues, $\lambda_i(GK)$, to ensure closed-loop stability. Further, note that these requirements are even more restrictive if the nominal plant has RHP poles or zeros. (See Maciejowski, 1989, for example, for an analysis of both SISO and MIMO loop shaping constraints for minimum phase and non minimum phase systems, and Freudenberg and Looze, 1987, where, for the SISO case, it is shown that performance and robust stability properties are limited by the presence of RHP poles and zeros.)

A loop-shaping approach that is somewhat simpler from the designer's point of view is that used in the Loop Transfer Recovery (LTR) method in LQG design. (See Kwakernaak and Sivan, 1972, and Doyle and Stein, 1979.) In this method, the designer specifies a desired singular value loop-shape, and the guaranteed stability properties of the LQG compensator ensure stability. The designer is not concerned with stability when generating the loop-shape, although the LQG requirements enforce a certain 'structure' on the final loop-shape. (For example, the loop-shape will always have a -20db/decade roll-off at high frequency.) LTR however, cannot systematically deal with plants with RHP zeros (see Stein and Athans, 1987), and is limited in that it can only guarantee performance and robust stability properties at either plant input, point (i), or plant output, point (ii), in Figure 6.1. This is a property of the loop shaping approach used. LTR enables the designer to specify, via loop shaping, *either* the closed-loop functions $(I - GK)^{-1}$, and $GK(I - GK)^{-1}$ associated with plant output, or functions $(I - KG)^{-1}$ and $(I - KG)^{-1}KG$ associated with plant input.

The design technique that is proposed in this chapter is similar in philosophy to LTR: The designer specifies a desired loop shape (ignoring closed-loop stability considerations at this stage), and then the 'shaped' plant is further compensated by a controller using the normalized LCF robust stabilization method of Chapter 4. In this case, the guaranteed stability properties of the H_∞ method ensure closed-loop stability. It will be shown however that the limitations of the LTR method do not apply for this approach.

6.3 The Design Procedure

We will now formally state the design procedure that was proposed in Section 6.1. The objective of this approach is to incorporate the simple performance/robustness trade-off obtained in loop shaping, with the guaranteed stability properties of H_∞ design methods.

'The Loop Shaping Design Procedure (LSDP)'

(1) Loop Shaping - Using a precompensator, W_1, and/or a postcompensator, W_2, the singular values of the nominal plant are 'shaped' to give a desired open-loop shape. The nominal plant, G, and 'shaping functions' W_1, W_2 are combined to form the 'shaped plant', G_S where $G_S = W_2 G W_1$. (See Figure 6.4a.) We assume that W_1 and W_2 are such that G_S contains no hidden unstable modes.

(2) Robust Stabilization - (a) Calculate ϵ_{max}. If $\epsilon_{max} \ll 1$ return to (1) and adjust W_1 and W_2. (b) Select $\epsilon \leq \epsilon_{max}$, then synthesize a feedback controller, K_∞, which robustly stabilizes the normalized left coprime factorization of G_S, with stability margin ϵ, using the approach outlined in Chapter 4. (See Figure 6.4b.)

(3) The final feedback controller, K, is then constructed by combining the H_∞ controller, K_∞, with the shaping functions, W_1 and W_2 such that (see Figure 6.4c)

$$K = W_1 K_\infty W_2.$$

Note that, in contrast to the classical loop shaping approach, the loop shaping here is done *without* explicit regard for the nominal plant phase information. That is, closed-loop stability requirements are disregarded at this stage. Also, the robust stabilization is done *without* frequency weighting. The parameter ϵ_{max} can be seen as an *indicator* of the success of the loop shaping, where we note by Theorem 4.3, $\epsilon_{max} < 1$ always. A small value of ϵ_{max} ($\epsilon_{max} \ll 1$) in Stage (2) always indicates incompatibility between the specified loop shape, the nominal plant phase, and robust closed-loop stability.

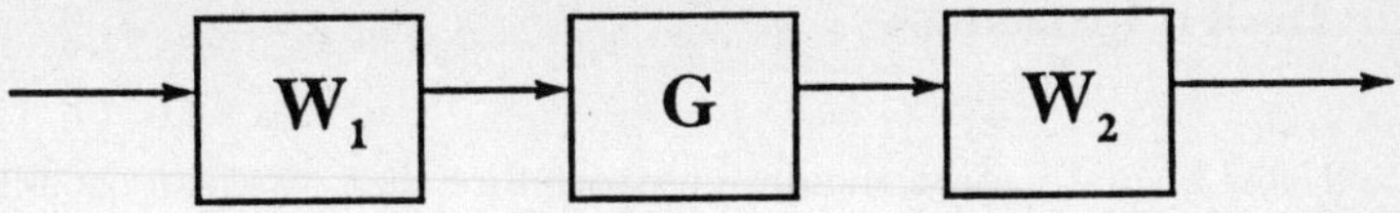

a) The Shaped Plant

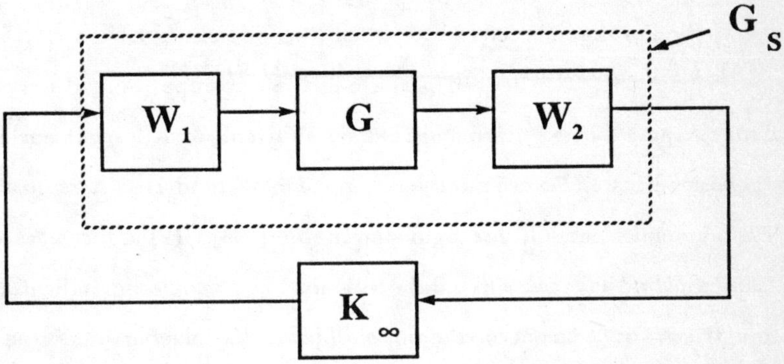

b) H_∞ Compensation

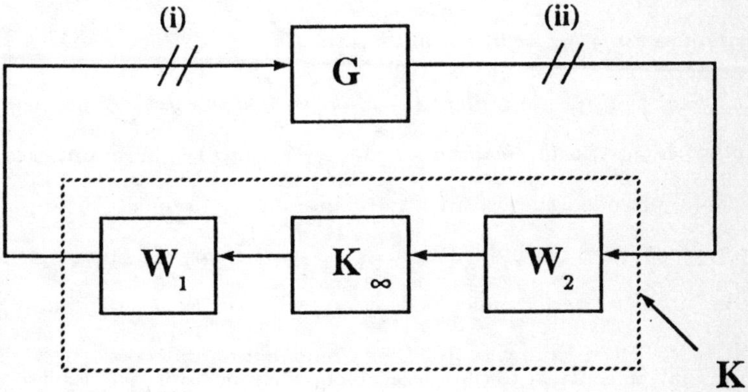

c) Final Controller

Figure 6.4 The Loop Shaping Design Procedure

Hence the loop shaping of Stage (1) is connected to the robust stabilization of Stage (2) via the indicator ϵ_{max}.

A typical design works as follows: The designer inspects the open-loop singular values of the nominal plant, and shapes these by pre and/or post compensation until nominal performance (and possibly robust stability) specifications are met. (Recall that the open-loop shape is related to closed-loop objectives.) Next ϵ_{max} is calculated. If ϵ_{max} is small, then the desired performance is incompatible with robust stability requirements, and the loop shape should be adjusted accordingly. For example, the desired bandwidth could be reduced by changes to $\boldsymbol{W}_1$ and $\boldsymbol{W}_2$. A feedback controller, $\boldsymbol{K}_\infty$, with associated stability margin (for the shaped plant) $\epsilon \leq \epsilon_{max}$ is then synthesized. Note that we have interpreted the measure ϵ_{max} as a design indicator rather than as a specific stability margin for the notional perturbations on the normalized LCF of the *shaped* plant. (We will justify this interpretation in Section 6.4.) This procedure is both simple and systematic, and only assumes knowledge of elementary loop shaping principles on the part of the designer.

Remark 6.1

The procedure is applicable to stable or unstable, minimum or non minimum phase plants, provided they satisfy the minimum requirement for any design – that is, there are no unstable hidden modes. In particular, if a plant is non-minimum phase, the standard performance restrictions will still exist in the design procedure: it is well known, (see Horowitz, 1963, or Freudenberg and Looze, 1987, for example) that the presence of a non minimum phase (RHP) zero in the nominal plant limits the achievable bandwidth and restricts low frequency behaviour. In the Loop Shaping Design Procedure, a loop shape which is incompatible with these general restrictions will lead to a 'small' ϵ_{max} value in Stage (2). To illustrate this, consider the following example:

Let a nominal plant be given by $\boldsymbol{G} = \frac{s-1}{s(s+2)}$, and we select $W = k$, a constant, as the shaping function. The shaped plant is then $\boldsymbol{G_S} = \boldsymbol{WG} = \frac{k(s-1)}{s(s+2)}$. By Horowitz,

(1963), we know that the closed-loop bandwidth has a practical limit of $1r/s$ because of the RHP zero at $s = 1$ in G. Noting that the crossover frequency of the open-loop shape reflects the closed-loop bandwidth, we now select four values of k, the largest yielding a crossover frequency that is incompatible with the bandwidth limitations. The resulting ϵ_{max} values after Stage (2) of the Loop Shaping Design Procedure are:

$$k = .1 \quad : \quad \epsilon_{max} = 0.6814$$

$$k = 1 \quad : \quad \epsilon_{max} = 0.5038$$

$$k = 5 \quad : \quad \epsilon_{max} = 0.2177$$

$$k = 20 \quad : \quad \epsilon_{max} = 0.0684$$

The value of ϵ_{max} associated with $k = 20$ is extremely small, indicating an 'incompatible loop shape'.

This incompatibility is confirmed by comparing the specified and achieved loop shapes for each case in Figure 6.5.

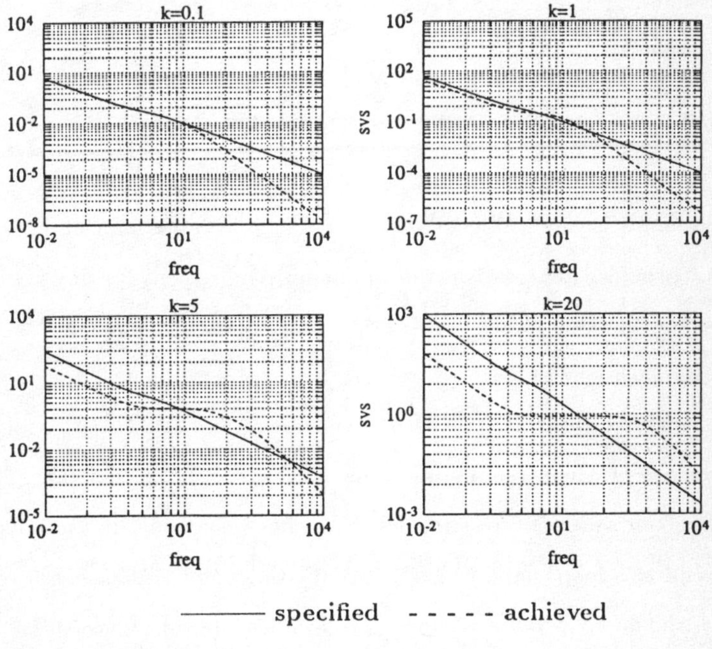

Figure 6.5 Specified and Achieved Loop Shapes for $k = 0.1, 1, 5, 20$

In the next section, we give theoretical justification for using the loop shaping design procedure.

6.4 Using ϵ as a Design Indicator

6.4.1 Guarantees on the Achieved Loop Shape

In the previous section we specified the desired loop shape by W_2GW_1 (Figure 6.4a), but the actual loop shape achieved is in fact given by $W_1K_\infty W_2G$ at plant input (point (i), Figure 6.4c), and $GW_1K_\infty W_2$ at plant output (point (ii), Figure 6.4c). We will show that the degradation in the loop shape caused by the H_∞ controller K_∞ is limited at those frequencies where the specified loop shape is sufficiently large or sufficiently small.

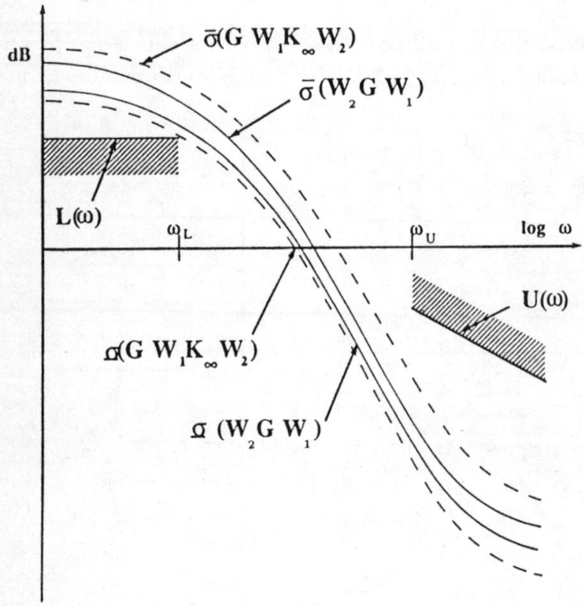

Figure 6.6 Specified and Achieved Loop Shapes

Figure 6.6 illustrates the discrepancies that may occur between specified and achieved loop shapes. It can be seen in Figure 6.6 that, at low frequency (in particular $\omega \in (0, \omega_l)$), the deterioration in loop shape at plant output can be obtained by comparing $\underline{\sigma}(GW_1K_\infty W_2)$ with $\underline{\sigma}(W_2GW_1)$. Note that:

$$\underline{\sigma}(GK) = \underline{\sigma}(GW_1K_\infty W_2) \geq \underline{\sigma}(W_2GW_1)\underline{\sigma}(K_\infty)/c(W_2), \tag{6.5}$$

where $c(\bullet)$ denotes the (frequency dependent) condition number. Similarly, for loop shape deterioration at plant input, we compare $\underline{\sigma}(W_1K_\infty W_2G)$ with $\underline{\sigma}(W_2GW_1)$ and we have

$$\underline{\sigma}(KG) = \underline{\sigma}(W_1K_\infty W_2G) \geq \underline{\sigma}(W_2GW_1)\underline{\sigma}(K_\infty)/c(W_1). \tag{6.6}$$

In each case, $\underline{\sigma}(K_\infty)$ is required to obtain a bound on the deterioration in the loop shape at low frequency. Note that the condition numbers $c(W_1)$ and $c(W_2)$ are selected by the designer, and are commonly of order one.

The following result shows that $\underline{\sigma}(K_\infty)$ is explicitly bounded by functions of ϵ and $\underline{\sigma}(G_S)$, the minimum singular value of the shaped plant. We assume here that the shaped plant G_S has an equal number of inputs and outputs. For convenience, the parameter, γ, defined $\gamma \triangleq \epsilon^{-1}$ will be used in the following analysis.

Theorem 6.2

Any controller, K_∞, *satisfying*

$$\left\| \begin{bmatrix} K_\infty \\ I \end{bmatrix} (I - G_S K_\infty)^{-1} \tilde{M}_s^{-1} \right\|_\infty \leq \gamma$$

where $(\tilde{N}_s, \tilde{M}_s)$ *is a normalized LCF of* G_S, *and* G_S *is assumed square, also satisfies*

$$\underline{\sigma}(K_\infty(j\omega)) \geq \frac{\underline{\sigma}(G_S(j\omega)) - (\gamma^2 - 1)^{1/2}}{(\gamma^2 - 1)^{1/2}\underline{\sigma}(G_S(j\omega)) + 1} \tag{6.7}$$

for all ω *such that*

$$\underline{\sigma}(G_S(j\omega)) > \sqrt{\gamma^2 - 1}.$$

Proof:

Firstly we note that

$$\left\| \begin{bmatrix} K_\infty \\ I \end{bmatrix} (I - G_S K_\infty)^{-1} \tilde{M}_s^{-1} \right\|_\infty \leq \gamma$$

can be rewritten as

$$(I + K_\infty^* K_\infty) \leq \gamma^2 (I - K_\infty^* G_S^*)(\tilde{M}_s^* \tilde{M}_s)(I - G_S K_\infty) \tag{6.8}$$

where the notation $A \leq B$ implies $(B - A) \geq 0$ in the usual way. Assuming that K_∞^{-1} exists, and noting that $\tilde{M}_s \tilde{M}_s^* = I - \tilde{N}_s \tilde{N}_s^*$, and $G_S = \tilde{M}_s^{-1} \tilde{N}_s$ then from (6.8) it follows that:

$$\begin{aligned} &(K_\infty^{*-1} K_\infty^{-1} + I) \leq \gamma^2 (K_\infty^{*-1} - G_S^*)(I + G_S G_S^*)^{-1}(K_\infty^{-1} - G_S) \\ \Rightarrow &K_\infty^{*-1}(I - \gamma^2 (I + G_S G_S^*)^{-1}) K_\infty^{-1} + \gamma^2 G_S^* (I + G_S G_S^*)^{-1} K_\infty^{-1} \\ &+ \gamma^2 K_\infty^{*-1}(I + G_S G_S^*)^{-1} G_S + I - \gamma^2 G_S^* (I + G_S G_S^*)^{-1} G_S \leq 0. \end{aligned} \tag{6.9}$$

(6.9) is a matrix quadratic function of K_∞^{-1}, and we note that the requirement

$$(I - \gamma^2 (I + G_S G_S^*)^{-1}) > 0 \tag{6.10}$$

is sufficient to enable 'completion of the square' in terms of K_∞^{-1}. Therefore, assuming that condition (6.10) holds, we define W such that

$$(W W^*)^{-1} \triangleq (I - \gamma^2 (I + G_S G_S^*)^{-1}), \tag{6.11}$$

and completing the square in (6.9) yields

$$(K_\infty^{*-1} + N^*)(W^{*-1} W^{-1})(K_\infty^{-1} + N) \leq (\gamma^2 - 1) R^* R. \tag{6.12}$$

N is defined by:

$$\begin{aligned} N &\triangleq \gamma^2 W W^* (I + G_S G_S^*)^{-1} G_S \\ &= \gamma^2 G_S ((1 - \gamma^2) I + G_S^* G_S)^{-1} \text{ by (6.11)} \end{aligned}$$

and R is defined such that

$$\begin{aligned} &(\gamma^2 - 1) R^* R \\ &= -I + \gamma^2 G_S^* (I + G_S G_S^*)^{-1} G_S + \gamma^4 G_S^* (I + G_S G_S^*)^{-1} W W^* (I + G_S G_S^*)^{-1} G_S \\ &= -I + \gamma^2 G_S^* G_S (I + G_S^* G_S)^{-1} + \gamma^4 G_S^* G_S (I + G_S^* G_S)^{-1} ((1 - \gamma^2) I + G_S^* G_S)^{-1} \\ &= (I + G_S^* G_S)^{-1} \{ -(I + G_S^* G_S)((1 - \gamma^2) I + G_S^* G_S) \\ &+ \gamma^2 G_S^* G_S ((1 - \gamma^2) I + G_S^* G_S) + \gamma^4 G_S^* G_S \} ((1 - \gamma^2) I + G_S^* G_S)^{-1} \\ &= (\gamma^2 - 1)(I + G_S^* G_S)^{-1} \{ I + 2 G_S^* G_S + (G_S^* G_S)^2 \} ((1 - \gamma^2) I + G_S^* G_S)^{-1} \\ &= (\gamma^2 - 1)(I + G_S^* G_S)((1 - \gamma^2) I + G_S^* G_S)^{-1}. \end{aligned} \tag{6.13}$$

Rearranging (6.12) we have the following condition on $\underline{\sigma}(K_\infty)$:

$$\begin{aligned} &\overline{\sigma}(W^{-1}(K_\infty^{-1}+N)R^{-1}) \le (\gamma^2-1)^{1/2} \\ &\quad\Rightarrow \overline{\sigma}(K_\infty^{-1}+N) \le (\gamma^2-1)^{1/2}\overline{\sigma}(W)\overline{\sigma}(R) \\ &\qquad\Rightarrow \underline{\sigma}(K_\infty) \ge \{(\gamma^2-1)^{1/2}\overline{\sigma}(W)\overline{\sigma}(R)+\overline{\sigma}(N)\}^{-1}. \qquad (6.14) \end{aligned}$$

To evaluate appropriate bounds on $\overline{\sigma}(W)$, $\overline{\sigma}(N)$ and $\overline{\sigma}(R)$ we first let G_S have a Singular Value Decomposition given by $G_S = Y\Sigma U^*$, where $\Sigma = \begin{bmatrix} \Sigma_1 & 0 \\ 0 & 0 \end{bmatrix}$ has the dimensions of the plant ($p = m$), $\Sigma_1 = diag(\sigma_1, \ldots, \sigma_r)$, $i = 1, \ldots, r$, $r \le m = p$, and Y and U are square invertible functions of appropriate dimension satisfying $YY^* = I$ and $UU^* = I$. Then,

$$\begin{aligned} WW^* &= (I+Y\Sigma^2Y^*)((1-\gamma^2)I+Y\Sigma^2Y^*)^{-1} \\ &= Y(I+\Sigma^2)((1-\gamma^2)I+\Sigma^2)^{-1}Y^* \end{aligned}$$

and

$$\sigma_i^2(W) = \lambda_i(WW^*) = \frac{1+\sigma_j^2(G_S)}{(1-\gamma^2)I+\sigma_j^2(G_S)}.$$

It is then straightforward to verify that

$$\overline{\sigma}(W) = \left\{\frac{1+\underline{\sigma}^2(G_S)}{\underline{\sigma}^2(G_S)-(\gamma^2-1)}\right\}^{1/2}. \qquad (6.15)$$

Note from (6.15) that the requirement $\underline{\sigma}(G_S) > (\gamma^2-1)^{1/2}$ is sufficient to guarantee that $(WW^*)^{-1} > 0$, and hence satisfy the assumption in (6.10).

Next note that, by inspection of (6.13) we have

$$\overline{\sigma}(R) = \overline{\sigma}(W) = \left\{\frac{1+\underline{\sigma}^2(G_S)}{\underline{\sigma}^2(G_S)-(\gamma^2-1)}\right\}^{1/2}. \qquad (6.16)$$

Finally, using a singular value decomposition approach again, it is straightforward to verify that

$$\overline{\sigma}(N) = \frac{\gamma^2\underline{\sigma}(G_S)}{\underline{\sigma}^2(G_S)-(\gamma^2-1)}. \qquad (6.17)$$

Using (6.15)-(6.17) in (6.14) we then have

$$\underline{\sigma}(K_\infty) \ge \left\{\frac{(\gamma^2-1)^{1/2}(1+\underline{\sigma}^2(G_S))+\gamma^2\underline{\sigma}(G_S)}{\underline{\sigma}^2(G_S)-(\gamma^2-1)}\right\}^{-1}$$

giving the required result after cancelling the factor $\underline{\sigma}(G_S)+\sqrt{\gamma^2-1}$. ■

The main implication of Theorem 6.2 is that the bound on $\underline{\sigma}(K_\infty)$ depends only on the selected loop shape, and the stability margin. The value of γ/ϵ directly determines the frequency range over which this result is valid - a small γ (large ϵ) is desirable, as we would expect. Further, if we consider those frequencies where $\underline{\sigma}(G_S) \gg \sqrt{\gamma^2 - 1}$, the following asymptotic result is immediate from Theorem 6.2:

Corollary 6.3

Following the notation of Theorem 6.2, if $\underline{\sigma}(G_S(j\omega)) \gg \sqrt{\gamma^2 - 1}$, *then*

$$\underline{\sigma}(K_\infty(j\omega)) \gtrsim \frac{1}{\sqrt{\gamma^2 - 1}},$$

where '$\gtrsim$' denotes 'asymptotically greater than or equal to as $\underline{\sigma}(G_S(j\omega)) \to \infty$*'.*

Further, referring again to the diagram in Figure 6.6, it can be seen that at high frequency (in particular $\omega \in (\omega_u, \infty)$) the deterioration in plant output loop shape can be obtained by comparing $\overline{\sigma}(GW_1K_\infty W_2)$ with $\overline{\sigma}(W_2GW_1)$. Note that, analogously to (6.5) and (6.6) we have

$$\overline{\sigma}(GK) = \overline{\sigma}(GW_1K_\infty W_2) \leq \overline{\sigma}(W_2GW_1)\overline{\sigma}(K_\infty)c(W_2). \tag{6.18}$$

Similarly, the corresponding deterioration in plant input loop shape is obtained by comparing $\overline{\sigma}(W_1K_\infty W_2G)$ with $\overline{\sigma}(W_2GW_1)$ where

$$\overline{\sigma}(KG) = \overline{\sigma}(W_1K_\infty W_2G) \leq \overline{\sigma}(W_2GW_1)\overline{\sigma}(K_\infty)c(W_1). \tag{6.19}$$

Hence in each case, $\overline{\sigma}(K_\infty)$ is required to obtain a bound on the deterioration in the loop shape at high frequency. In an analogous manner to Theorem 6.2, we now show that $\overline{\sigma}(K_\infty)$ is explicitly bounded by functions of γ, and $\overline{\sigma}(G_S)$, the maximum singular value of the shaped plant.

Theorem 6.4

Any controller, K_∞, *satisfying*

$$\left\| \begin{bmatrix} K_\infty \\ I \end{bmatrix} (I - G_S K_\infty)^{-1} \tilde{M}_s^{-1} \right\|_\infty \leq \gamma$$

where $(\tilde{N}_s, \tilde{M}_s)$ is a normalized LCF of G_S, also satisfies

$$\overline{\sigma}(K_\infty(j\omega)) \leq \frac{(\gamma^2-1)^{1/2} + \overline{\sigma}(G_S(j\omega))}{1-(\gamma^2-1)^{1/2}\overline{\sigma}(G_S(j\omega))} \tag{6.20}$$

for all ω such that

$$\overline{\sigma}(G_S(j\omega)) < \frac{1}{\sqrt{\gamma^2-1}}.$$

Proof:

The proof of Theorem 6.4 is similar to that of Theorem 6.2, and is only sketched here:

Noting that $\tilde{M}_s{}^* \tilde{M}_s = (I + G_S G_S^*)^{-1}$ we have by (6.8) that

$$\begin{aligned}
&(I + K_\infty^* K_\infty) \leq \gamma^2 (I - K_\infty^* G_S^*)(I + G_S G_S^*)^{-1}(I - G_S K_\infty) \\
&\Rightarrow K_\infty^*(I - \gamma^2 G_S^*(I + G_S G_S{}^*)^{-1} G_S) K_\infty + \gamma^2 K_\infty^* G_S^*(I + G_S G_S{}^*)^{-1} \\
&+ \gamma^2 (I + G_S G_S{}^*)^{-1} G_S K_\infty + I - \gamma^2 (I + G_S G_S{}^*)^{-1} \leq 0.
\end{aligned} \tag{6.21}$$

We again complete the square, this in terms of K_∞, noting that

$$(I - \gamma^2 G_S^*(I + G_S G_S{}^*)^{-1} G_S) > 0 \tag{6.22}$$

is a sufficient requirement. Defining function V such that

$$V^* V \triangleq (I - \gamma^2 G_S^*(I + G_S G_S{}^*)^{-1} G_S). \tag{6.23}$$

then by (6.21) we then have

$$(K_\infty^* + M^*)(V^* V)(K_\infty + M) \leq (\gamma^2 - 1) Y^* Y, \tag{6.24}$$

where

$$\begin{aligned}
M^* &= \gamma^2 (I + G_S G_S{}^*)^{-1} G_S V^{-1} V^{*-1} \\
&= \gamma^2 (I + (1-\gamma^2) G_S G_S{}^*)^{-1} G_S,
\end{aligned} \tag{6.25}$$

and Y is such that

$$
\begin{aligned}
&(\gamma^2-1)Y^*Y \\
&= -I+\gamma^2(I+G_SG_S{}^*)^{-1}+\gamma^4(I+G_SG_S{}^*)^{-1}G_SV^{-1}V^{*-1}G_S^*(I+G_SG_S{}^*)^{-1} \\
&= (I+G_SG_S{}^*)^{-1}\{-(I+G_SG_S{}^*)(I+(1-\gamma^2)G_SG_S{}^*) \\
&+\gamma^2(I+(1-\gamma^2)G_SG_S{}^*)+\gamma^4G_SG_S{}^*\}(I+(1-\gamma^2)G_SG_S{}^*)^{-1} \\
&= (\gamma^2-1)(I+G_SG_S{}^*)(I+(1-\gamma^2)G_SG_S{}^*)^{-1}. \qquad (6.26)
\end{aligned}
$$

By (6.24) and (6.26) we then have that

$$
\begin{aligned}
&\overline{\sigma}(V(K_\infty+M)Y^{-1}) \le (\gamma^2-1)^{1/2} \\
&\Rightarrow \overline{\sigma}(K_\infty) \le \frac{(\gamma^2-1)^{1/2}}{\underline{\sigma}(V)\underline{\sigma}(Y^{-1})}+\overline{\sigma}(M). \qquad (6.27)
\end{aligned}
$$

Using the singular value decomposition of G_S as in the proof of Theorem 6.2,

$$\sigma_i^2(Y^{-1}) = \sigma_i^2(V) = \lambda_i(V^*V) = \frac{1-(\gamma^2-1)\lambda_j(G_S{}^*G_S)}{1+\lambda_j(G_S{}^*G_S)}$$

and it is then straightforward to show

$$\underline{\sigma}(Y^{-1}) = \underline{\sigma}(V) = \left\{\frac{1-(\gamma^2-1)\overline{\sigma}^2(G_S)}{1+\overline{\sigma}^2(G_S)}\right\}^{1/2}. \qquad (6.28)$$

Note by (6.22) that $\overline{\sigma}(G_S) \le \frac{1}{(\gamma^2-1)^{1/2}}$ guarantees that $V^*V > 0$, so that the results of this theorem hold for all frequencies satisfying $\overline{\sigma}(G_S) \le \frac{1}{(\gamma^2-1)^{1/2}}$. Next, using the singular value decomposition approach again, we have that

$$\overline{\sigma}(M) = \frac{\gamma^2\overline{\sigma}(G_S)}{1-(\gamma^2-1)\overline{\sigma}^2(G_S)}, \qquad (6.29)$$

and finally using (6.28) and (6.29) in (6.27) we have

$$\overline{\sigma}(K_\infty) \le \frac{(\gamma^2-1)^{1/2}(1+\overline{\sigma}^2(G_S))}{1-(\gamma^2-1)\overline{\sigma}^2(G_S)}+\frac{\gamma^2\overline{\sigma}(G_S)}{1-(\gamma^2-1)\overline{\sigma}^2(G_S)}$$

as required. ■

Again, if we consider those frequencies at which $\overline{\sigma}(G_S) \ll \frac{1}{\sqrt{\gamma^2-1}}$, then the following asymptotic result is immediate from Theorem 6.4.

Corollary 6.5

Following the notation of Theorem 6.4, if $\overline{\sigma}(G_S(j\omega)) \ll \frac{1}{\sqrt{\gamma^2-1}}$*, then*

$$\overline{\sigma}(K_\infty(j\omega)) \lesssim \sqrt{\gamma^2 - 1}, \tag{6.30}$$

where '$\lesssim$*' denotes 'asymptotically less than or equal to as* $\overline{\sigma}(G_S(j\omega)) \to 0$*'.*

Remark 6.6

The approximate results in Corollaries 6.3 and 6.5 show that at frequencies where $\underline{\sigma}(G_S) \gg 1$ or $\overline{\sigma}(G_S) \ll 1$, the deterioration in the loop shape due to K_∞ is bounded by a function of γ (or equivalently ϵ) only. Noting that γ 'small' (implying ϵ 'large') indicates a minimal deterioration in the loop shape, it is confirmed that ϵ indicates the compatibility between the specified loop shape and closed-loop stability requirements.

Remark 6.7

To show that the expression $(\gamma^2 - 1)^{1/2}$ always exists, recall that by Theorem 4.3,

$$\gamma = \epsilon^{-1} \geq \{1 - \|[\tilde{N}, \tilde{M}]\|_H^2\}^{-1/2}.$$

Noting that $\|[\tilde{N}, \tilde{M}]\|_H > 0$ we have that

$$\gamma > 1 \Rightarrow (\gamma^2 - 1) > 0 \tag{6.31}$$

and hence $(\gamma^2 - 1)^{1/2}$ always exists.

Remark 6.8

Note that in Theorem 6.4 it is not necessary to assume that G_S is a square transfer function matrix. Hence, high frequency loop shape properties can be guaranteed for a plant of any dimension.

Remark 6.9

In Section 4.5.2 we showed that the class of positive real systems always satisfies the property

$$\gamma_{min} = \epsilon_{max}^{-1} \leq \sqrt{2},$$

where ϵ_{max}, the maximum stability margin, is achieved by the optimal normalized LCF controller. Assuming in this case that K_∞ gives $\gamma \leq \sqrt{2}$, then in Theorem 6.2, this implies that for a positive real G_S we have

$$\underline{\sigma}(K_\infty) \geq \frac{\underline{\sigma}(G_S) - 1}{\underline{\sigma}(G_S) + 1}$$

for all frequencies such that $\underline{\sigma}(G_S) > 1$. Similarly, in Theorem 6.4, G_S positive real implies

$$\overline{\sigma}(K_\infty) \leq \frac{\overline{\sigma}(G_S) + 1}{1 - \overline{\sigma}(G_S)}$$

for all frequencies such that $\overline{\sigma}(G_S) < 1$. Further note from the above results that at those frequencies where $\underline{\sigma}(G_S) \gg 1$ we have $\underline{\sigma}(K_\infty) \geq 1$ and at frequencies where $\overline{\sigma}(G_S) \ll 1$ we have $\overline{\sigma}(K_\infty) \leq 1$. Hence, there will be *no* deterioration in the loop shape at high gain (respectively low gain) due to the inclusion of K_∞.

We have shown in this section that the values of γ (alternatively ϵ) achieved in the Loop Shaping Design Procedure will directly affect the singular values of the H_∞ controller K_∞. Although the precise relationship between γ and the shaping functions W_1 and W_2 is not known, we have shown that a large γ value indicates incompatibility between the selected loop shape and the closed-loop stability requirements. Such an incompatibility can lead to extensive deterioration in the specified loop shape. Under such circumstances, the designer would be required to 'relax' the original loop shape specification, until a more compatible loop shape is achieved.

6.4.2 Assessing Stabilizing Controllers using ϵ

In Section 6.4.1 we evaluated bounds on the *maximum* loop shape deterioration at low and high loop gains, for a given $\epsilon(=\gamma^{-1})$, showing that the achieved loop shape can differ from the specified loop shape by only a limited amount. This, combined with the robust stability guarantees implied by ϵ indicates that a controller achieving a *large* ϵ will guarantee that the final design has good robust stability properties (in a coprime factor sense) and that a loop shape 'close' to the specified loop shape has been achieved. In this section we take a contrary approach: for a controller achieving a *small* stability margin, that is $\epsilon \ll 1$, is it *still* possible to guarantee that the final design has good robust stability properties and that a loop shape 'close' to the specified loop shape has been achieved? To show that this is not the case, we note that, given any stabilizing controller, K_a, there exists a frequency ω_o, such that

$$\overline{\sigma}\left(\begin{bmatrix} K_a \\ I \end{bmatrix}(I - G_S K_a)^{-1}\tilde{M}_s^{-1}\right)(j\omega_o) = \gamma \geq \gamma_{min}.$$

We can now evaluate the *minimum* loop shape deterioration that can be expected, if any, for this γ, and examine the effect of γ on robust stability at this frequency. These ideas are summarized in the following theorem.

Theorem 6.10

Let $\epsilon_{max} = \gamma_{min}^{-1}$ be the optimal stability margin for the normalized LCF robust stabilization problem as given in Theorem 4.3. Then, for any stabilizing controller, K_a, there exists a frequency, ω_o, such that

$$\overline{\sigma}\left(\begin{bmatrix} K_a \\ I \end{bmatrix}(I - G_S K_a)^{-1}\tilde{M}_s^{-1}\right)(j\omega_o) \geq \gamma_{min}, \tag{6.32}$$

and hence

(i) there exists a perturbed system

$$G_\Delta = W_2^{-1}(\tilde{M}_s + \Delta_M)^{-1}(\tilde{N}_s + \Delta_N)W_1^{-1} \tag{6.33}$$

with

$$\|[\Delta_N, \Delta_M]\|_\infty \leq \gamma_{min}^{-1} \tag{6.34}$$

which destabilizes the system. Further,

(ii) if $\underline{\sigma}(G_S(j\omega_o)) > (\gamma_{min}^2 - 1)^{-1/2}$, *then*

$$\underline{\sigma}(K_a(j\omega_o)) \leq \frac{(\gamma_{min}^2 - 1)^{1/2} + \underline{\sigma}(G_S(j\omega_o))}{(\gamma_{min}^2 - 1)^{1/2}\underline{\sigma}(G_S(j\omega_o)) - 1}, \tag{6.35}$$

and

(iii) if $\overline{\sigma}(G_S(j\omega_o)) < (\gamma_{min}^2 - 1)^{1/2}$, *then*

$$\overline{\sigma}(K_a(j\omega_o)) \geq \frac{(\gamma_{min}^2 - 1)^{1/2} - \overline{\sigma}(G_S(j\omega_o))}{(\gamma_{min}^2 - 1)^{1/2}\overline{\sigma}(G_S(j\omega_o)) + 1} \tag{6.36}$$

Proof:

The proof of (i) follows immediately by noting Corollary 3.7.

The proofs for (ii) and (iii) are, in the main part, analogous to the proofs of Theorems 6.2 and 6.4. The procedure is therefore only sketched here.

To prove (ii) we note by (6.32) that, at frequency ω_o, there exists a vector, x, such that

$$x^*(I + K_a^* K_a)x \geq \gamma_{min}^2 x^*(I - K_a^* G_S^*)(I + G_S G_S^*)^{-1}(I - G_S K_a)x \tag{6.37}$$

$$\Rightarrow x^*\{K_a^*(I - \gamma_{min}^2 G_S^*(I + G_S G_S{}^*)^{-1} G_S)K_a + \gamma_{min}^2 K_a^* G_S^*(I + G_S G_S{}^*)^{-1}$$

$$+ \gamma_{min}^2 (I + G_S G_S{}^*)^{-1} G_S K_a + I - \gamma_{min}^2 (I + G_S G_S{}^*)^{-1}\}x \geq 0. \tag{6.38}$$

We again complete the squares in terms of K_a, noting that

$$(I - \gamma_{min}^2 G_S^*(I + G_S G_S{}^*)^{-1} G_S) < 0 \tag{6.39}$$

is a sufficient requirement, and noting further that this is satisfied if

$$\underline{\sigma}(G_S) > (\gamma_{min}^2 - 1)^{-1/2}. \tag{6.40}$$

Analogously to the proof of Theorem 6.4, we then have, from (6.38),

$$\boldsymbol{x}^*\{(\boldsymbol{K}_a^* - \boldsymbol{M}_1^*)(\boldsymbol{V}_1^{*-1}\boldsymbol{V}_1^{-1})(\boldsymbol{K}_a - \boldsymbol{M}_1)\}\boldsymbol{x} \leq (\gamma_{min}^2 - 1)\boldsymbol{x}^*\boldsymbol{Y}_1^*\boldsymbol{Y}_1\boldsymbol{x} \tag{6.41}$$

where $\boldsymbol{V}_1$, $\boldsymbol{M}_1$ and $\boldsymbol{W}_1$ satisfy

$$-(\boldsymbol{V}_1^{*-1}\boldsymbol{V}_1^{-1}) = (I - \gamma_{min}^2 \boldsymbol{G}_S^*(I + \boldsymbol{G}_S\boldsymbol{G}_S{}^*)^{-1}\boldsymbol{G}_S) \tag{6.42}$$

$$\boldsymbol{Y}_1^*\boldsymbol{Y}_1 = (\boldsymbol{G}_S\boldsymbol{G}_S^* + I)((\gamma_{min}^2 - 1)\boldsymbol{G}_S\boldsymbol{G}_S^* - I)^{-1} \tag{6.43}$$

$$\boldsymbol{M}_1 = \gamma_{min}^2 \boldsymbol{V}_1\boldsymbol{V}_1^*\boldsymbol{G}_S^*(I + \boldsymbol{G}_S\boldsymbol{G}_S{}^*)^{-1} \tag{6.44}$$

By (6.41) we then have that

$$\begin{aligned}
&\|\boldsymbol{V}_1^{-1}(\boldsymbol{K}_a - \boldsymbol{M}_1)\boldsymbol{x}\|_2 \leq (\gamma_{min}^2 - 1)^{1/2}\|\boldsymbol{Y}_1\boldsymbol{x}\|_2 \\
&\Rightarrow \underline{\sigma}(\boldsymbol{V}_1^{-1})\underline{\sigma}(\boldsymbol{K}_a - \boldsymbol{M}_1) \leq (\gamma_{min}^2 - 1)^{1/2}\overline{\sigma}(\boldsymbol{Y}_1) \\
&\qquad \Rightarrow \underline{\sigma}(\boldsymbol{K}_a) \leq (\gamma_{min}^2 - 1)^{1/2}\overline{\sigma}(\boldsymbol{Y}_1)\overline{\sigma}(\boldsymbol{V}_1) + \overline{\sigma}(\boldsymbol{M}_1).
\end{aligned} \tag{6.45}$$

It is then simple to verify that

$$\overline{\sigma}(\boldsymbol{V}_1) = \overline{\sigma}(\boldsymbol{Y}_1) = \left\{\frac{1 + \underline{\sigma}^2(\boldsymbol{G}_S)}{(\gamma_{min}^2 - 1)\underline{\sigma}^2(\boldsymbol{G}_S) - 1}\right\}^{1/2} \tag{6.46}$$

and

$$\overline{\sigma}(\boldsymbol{M}_1) = \frac{\gamma_{min}^2\underline{\sigma}(\boldsymbol{G}_S)}{(\gamma_{min}^2 - 1)\underline{\sigma}^2(\boldsymbol{G}_S) - 1}, \tag{6.47}$$

and (6.46) and (6.47) in (6.45) then completes the proof of (ii).

The proof of (iii) closely follows the proof of Theorem 6.2, and is outlined below.

By (6.37), at frequency ω_o, there exists a vector $\boldsymbol{y}$, $\boldsymbol{y} \triangleq \boldsymbol{K}_a\boldsymbol{x}$, such that

$$\begin{aligned}
&\boldsymbol{y}^*(\boldsymbol{K}_a^{*-1}\boldsymbol{K}_a^{-1} + I)\boldsymbol{y} \geq \boldsymbol{y}^*\gamma_{min}^2(\boldsymbol{K}_a^{*-1} - \boldsymbol{G}_S^*)(I + \boldsymbol{G}_S\boldsymbol{G}_S{}^*)^{-1}(\boldsymbol{K}_a^{-1} - \boldsymbol{G}_S)\boldsymbol{y} \\
&\Rightarrow \boldsymbol{y}^*\{\boldsymbol{K}_a^{*-1}(I - \gamma_{min}^2(I + \boldsymbol{G}_S\boldsymbol{G}_S{}^*)^{-1})\boldsymbol{K}_a^{-1} + \gamma_{min}^2\boldsymbol{G}_S^*(I + \boldsymbol{G}_S\boldsymbol{G}_S{}^*)^{-1}\boldsymbol{K}_a^{-1} \\
&\quad + \gamma_{min}^2\boldsymbol{K}_a^{*-1}(I + \boldsymbol{G}_S\boldsymbol{G}_S{}^*)^{-1}\boldsymbol{G}_S + I - \gamma_{min}^2\boldsymbol{G}_S^*(I + \boldsymbol{G}_S\boldsymbol{G}_S{}^*)^{-1}\boldsymbol{G}_S\}\boldsymbol{y} \geq 0.
\end{aligned} \tag{6.48}$$

Assuming that

$$(I - \gamma_{min}^2 (I + G_S G_S{}^*)^{-1}) < 0,$$

for which

$$\overline{\sigma}(G_S(j\omega_o)) < (\gamma_{min}^2 - 1)^{1/2} \tag{6.49}$$

is a sufficient condition, we 'complete of the square' in (6.48) in terms of K_a^{-1}. Analogously with the proof of Theorem 6.2, we then have

$$y^*\{(K_a^{*-1} - N_1^*)(W_1^{*-1} W_1^{-1})(K_a^{-1} - N_1)\}y \leq (\gamma_{min}^2 - 1) y^* R_1^* R_1 y \tag{6.50}$$

where W_1, R_1 and N_1 satisfy

$$\begin{aligned} -(W_1 W_1{}^*)^{-1} &= (I - \gamma_{min}^2 (I + G_S G_S{}^*)^{-1}) \\ R_1^* R_1 &= (I + G_S{}^* G_S)((\gamma_{min}^2 - 1)I - G_S{}^* G_S)^{-1} \\ N_1 &= \gamma_{min}^2 G_S((\gamma_{min}^2 - 1)I - G_S{}^* G_S)^{-1}. \end{aligned}$$

By (6.50) we can show

$$\underline{\sigma}(K_a) \geq (\overline{\sigma}(R_1)\overline{\sigma}(W_1) + \overline{\sigma}(N_1))^{-1}. \tag{6.51}$$

It is then straightforward to show (using the usual singular value decomposition approach) that

$$\overline{\sigma}(R_1) = \overline{\sigma}(W_1) = \left\{ \frac{1 + \overline{\sigma}^2(G_S)}{\gamma_{min}^2 - 1 - \overline{\sigma}^2(G_S)} \right\}^{1/2}$$

and

$$\overline{\sigma}(N_1) = \frac{\gamma_{min}^2 \overline{\sigma}(G_S)}{\gamma_{min}^2 - 1 - \overline{\sigma}^2(G_S)},$$

and substitution of these results in (6.51) gives the desired result. ■

Remark 6.11

Theorem 6.10 states that, for any $\epsilon \leq \epsilon_{max}$, there will be a *minimum* deterioration in the desired loop shape at frequencies of high or low loop gain or there will exist a perturbation satisfying (6.34) which destabilizes the closed-loop system. This result shows that γ/ϵ is a design indicator for the entire frequency range: Firstly, if the coprime factor robust stability is poor (that is $\epsilon \ll 1$ or $\gamma \gg 1$) at a frequency where $\tilde{N}_s$ and $\tilde{M}_s$ are of comparable size, then (i) shows that only a small relative perturbation on either $\tilde{N}_s$ or $\tilde{M}_s$ is permitted. However, if $\tilde{M}_s \gg \tilde{N}_s$, then the notional destabilizing perturbation could correspond to a very large relative perturbation in $\tilde{N}_s$, implying that robust stability properties may in fact be acceptable despite $\gamma \gg 1$. However, in this case, we have $\overline{\sigma}(G_S) \ll 1$ and part (iii) of Theorem 6.10 applies, showing that the loop gain is necessarily and undesirably increased. Conversely, if $\tilde{M}_s \ll \tilde{N}_s$, then the notional destabilizing perturbation could correspond to a very large relative perturbation in $\tilde{M}_s$, again implying that robust stability properties may in fact be acceptable despite $\gamma \gg 1$. However, in this case, we have $\underline{\sigma}(G_S) \gg 1$ and part (ii) of Theorem 6.10 applies, showing that the loop gain is necessarily and undesirably decreased.

The results in this section have shown that ϵ is an effective design indicator for the Loop Shaping Design Procedure, and have confirmed that the normalized LCF robust stabilization design approach provides a suitable stabilizing controller. In the next section we examine the closed-loop behaviour achieved using the Loop Shaping Design Procedure, and show that the loop shaping approach gives guarantees on particular closed-loop objectives.

6.5 Shaping Functions and Closed-Loop Behaviour

In this section we examine the effect of the shaping functions, W_1 and W_2, on closed-loop behaviour.

6.5.1 Behaviour of Standard Closed-Loop Objectives

In Section 6.2 we stated that a feature of the classical loop shaping design approach is that it is possible to ensure, by open-loop shaping, that a number of standard closed-loop design objectives are bounded. The following result demonstrates that this is also the case for the Loop Shaping Design Procedure outlined in Section 6.3:

Theorem 6.12

Let G be the nominal plant and let $K = W_1 K_\infty W_2$ be the associated controller obtained from the Loop Shaping Design Procedure of Section 6.3. Then if

$$\left\| \begin{bmatrix} K_\infty \\ I \end{bmatrix} (I - G_S K_\infty)^{-1} \tilde{M}_s^{-1} \right\|_\infty \leq \gamma$$

we have

$$\overline{\sigma}(K(I - GK)^{-1}) \leq \gamma \overline{\sigma}(\tilde{M}_s) \overline{\sigma}(W_1) \overline{\sigma}(W_2) \tag{6.52}$$

$$\overline{\sigma}((I - GK)^{-1}) \leq \gamma \overline{\sigma}(\tilde{M}_s) c(W_2) \tag{6.53}$$

$$\overline{\sigma}(K(I - GK)^{-1} G) \leq \gamma \overline{\sigma}(\tilde{N}_s) c(W_1) \tag{6.54}$$

$$\overline{\sigma}((I - GK)^{-1} G) \leq \frac{\gamma \overline{\sigma}(\tilde{N}_s)}{\underline{\sigma}(W_1) \underline{\sigma}(W_2)} \tag{6.55}$$

and

$$\overline{\sigma}((I - KG)^{-1}) \leq 1 + \gamma \overline{\sigma}(\tilde{N}_s) c(W_1) \tag{6.56}$$

$$\overline{\sigma}(G(I - KG)^{-1} K) \leq 1 + \gamma \overline{\sigma}(\tilde{M}_s) c(W_2) \tag{6.57}$$

where $(\tilde{N}_s, \tilde{M}_s)$ is a normalized LCF of $G_S = W_2 G W_1$, and $c(\bullet)$ denotes $\overline{\sigma}(\bullet)/\underline{\sigma}(\bullet)$, the (frequency dependent) condition number.

Proof:

Firstly note that

$$\left\| \begin{bmatrix} K_\infty \\ I \end{bmatrix} (I - G_S K_\infty)^{-1} \tilde{M}_s^{-1} \right\|_\infty \leq \gamma$$

$$\Rightarrow \left\| \begin{bmatrix} W_1{}^{-1} K \\ W_2 \end{bmatrix} (I - GK)^{-1} W_2{}^{-1} \tilde{M}_s^{-1} \right\|_\infty \leq \gamma \tag{6.58}$$

Noting that $\overline{\sigma}(\bullet) \leq \| \bullet \|_\infty$ for all frequencies, we can immediately show (6.52) and (6.53), because by (6.58) we have

$$\overline{\sigma}(W_1{}^{-1} K (I - GK)^{-1} W_2{}^{-1} \tilde{M}_s^{-1}) \leq \gamma \tag{6.59}$$

$$\Rightarrow \overline{\sigma}(K(I - GK)^{-1}) \leq \gamma \overline{\sigma}(\tilde{M}_s) \overline{\sigma}(W_1) \overline{\sigma}(W_2)$$

and

$$\overline{\sigma}(W_2 (I - GK)^{-1} W_2{}^{-1} \tilde{M}_s^{-1}) \leq \gamma \tag{6.60}$$

$$\Rightarrow \overline{\sigma}((I - GK)^{-1}) \leq \gamma \overline{\sigma}(\tilde{M}_s) c(W_2).$$

Next recall that $G_S = W_2 G W_1 = \tilde{M}_s^{-1} \tilde{N}_s$. Then by (6.59) we have

$$\overline{\sigma}(W_1{}^{-1} K (I - GK)^{-1} G W_1) \leq \gamma \overline{\sigma}(\tilde{N}_s)$$

$$\Rightarrow \overline{\sigma}(K(I - GK)^{-1} G) \leq \gamma \overline{\sigma}(\tilde{N}_s) c(W_1),$$

and by (6.60)

$$\overline{\sigma}(W_2 (I - GK)^{-1} G W_1) \leq \gamma \overline{\sigma}(\tilde{N}_s)$$

$$\Rightarrow \overline{\sigma}((I - GK)^{-1} G) \leq \frac{\gamma \overline{\sigma}(\tilde{N}_s)}{\underline{\sigma}(W_1) \underline{\sigma}(W_2)}.$$

The results in (6.56) and (6.57) follow immediately from those in (6.54) and (6.53) by noting

$$\overline{\sigma}((I - KG)^{-1}) = \overline{\sigma}(I + K(I - GK)^{-1} G)$$

$$\leq 1 + \overline{\sigma}(K(I - GK)^{-1} G)$$

and

$$\overline{\sigma}(G(I - KG)^{-1} K) = \overline{\sigma}(I - (I - GK)^{-1})$$

$$\leq 1 + \overline{\sigma}((I - GK)^{-1}).$$

■

Remark 6.13

Note that, because $(\tilde{N}_s, \tilde{M}_s)$ is a *normalized* LCF of G_S, we have that $\overline{\sigma}(\tilde{N}_s) \leq 1$ and $\overline{\sigma}(\tilde{M}_s) \leq 1$ for all frequencies. Noting also that γ has a finite value (typically $2 < \gamma < 10$ in practice), and that the shaping functions are selected by the designer, then it can be seen that, by (6.52) - (6.57), all of the closed-loop objectives are guaranteed to have bounded magnitude. We say that in this case the objectives are *well behaved*.

For the simple case of $W_1 = I$ and $W_2 = I$ it can be readily seen that in (6.52) - (6.55), $\tilde{N}_s$ $(= \tilde{N})$ and $\tilde{M}_s$ $(= \tilde{M})$ provide frequency weighting. The frequency shapes are 'natural' for many problems: for example at frequencies where $\overline{\sigma}(\tilde{M}_s)$ is small (indicating a nominal plant pole near the imaginary axis), then $\overline{\sigma}((I - GK)^{-1})$ and $\overline{\sigma}(K(I - GK)^{-1})$ are small. So the behaviour of the standard closed-loop transfer functions $(I - GK)^{-1}$, $K(I - GK)^{-1}$, $K(I - GK)^{-1}G$, and $(I - GK)^{-1}G$ is compatible with sensible closed-loop design.

Remark 6.14

It is interesting to compare the guarantees on the closed-loop objectives in Theorem 6.12 with those obtained from another H_∞ optimization problem: Consider the H_∞ Weighted Mixed Sensitivity problem, where a stabilizing controller, K, is found such that

$$\left\| \begin{bmatrix} W_1(I - GK)^{-1} \\ W_2GK(I - GK)^{-1} \end{bmatrix} \right\|_\infty \leq \gamma \tag{6.61}$$

where W_1 and W_2 are stable, minimum phase functions. By (6.61) we then have

$$\overline{\sigma}(K(I - GK)^{-1}) \leq \gamma \overline{\sigma}(K)/\underline{\sigma}(W_1) \tag{6.62}$$

$$\overline{\sigma}((I - GK)^{-1}) \leq \gamma/\underline{\sigma}(W_1) \tag{6.63}$$

$$\overline{\sigma}(GK(I - GK)^{-1}) \leq \gamma/\underline{\sigma}(W_2) \tag{6.64}$$

$$\overline{\sigma}((I - GK)^{-1}G) \leq \gamma \overline{\sigma}(G)/\underline{\sigma}(W_1) \tag{6.65}$$

and assuming G and K are square

$$\overline{\sigma}((I - KG)^{-1}) \leq \gamma c(G)/\underline{\sigma}(W_1) \tag{6.66}$$

$$\overline{\sigma}(KG(I - KG)^{-1}) \leq \gamma c(K)/\underline{\sigma}(W_1). \tag{6.67}$$

Hence while (6.63) and (6.64) are well behaved, the remaining four objectives could be arbitrarily large, depending on the singular values and condition number of the nominal plant and resulting controller. In particular, note that (6.66) and (6.67), measures of performance and robustness respectively at the plant input, can be arbitrarily worse (depending on condition numbers $c(G)$ and $c(K)$) than the corresponding measures at plant output. In contrast, the normalized LCF design method guarantees good behaviour at both plant input and output.

6.5.2 Bounds on the Normalized Coprime Factors

In Theorem 6.12 we showed that the behaviour of a number of different closed-loop transfer functions is dependent on the normalized LCF of G_S, $(\tilde{N}_s, \tilde{M}_s)$. In Remark 6.13 we noted that $\overline{\sigma}(\tilde{N}_s) \leq 1$ and $\overline{\sigma}(\tilde{M}_s) \leq 1$, and we now state a technical result which demonstrates that $\overline{\sigma}(\tilde{N}_s)$ and $\overline{\sigma}(\tilde{M}_s)$ are related to the nominal plant G and the shaping functions W_1 and W_2.

Lemma 6.15

Let the shaped plant, $G_S = W_2 G W_1$, have a normalized LCF given by $(\tilde{N}_s, \tilde{M}_s)$. Then

$$\overline{\sigma}(\tilde{N}_s) = \left(\frac{\overline{\sigma}^2(W_2 G W_1)}{1 + \overline{\sigma}^2(W_2 G W_1)}\right)^{1/2} \tag{6.68}$$

and

$$\overline{\sigma}(\tilde{M}_s) = \left(\frac{1}{1 + \underline{\sigma}^2(W_2 G W_1)}\right)^{1/2}. \tag{6.69}$$

Proof:

Note, that by the definition of the normalized LCF in (4.1) we have

$$\begin{aligned} \tilde{M}_s \tilde{M}_s{}^* &= I - \tilde{N}_s \tilde{N}_s{}^* \\ &= I - \tilde{M}_s W_2 G W_1 W_1{}^* G^* W_2{}^* \tilde{M}_s{}^* \end{aligned}$$

$$\Rightarrow \tilde{M}_s{}^* \tilde{M}_s = (I + W_2 G W_1 W_1{}^* G^* W_2{}^*)^{-1}, \tag{6.70}$$

and similarly,

$$\begin{aligned} \tilde{N}_s{}^* \tilde{N}_s &= G_S{}^* \tilde{M}_s{}^* \tilde{M}_s G_S \\ &= (W_1{}^* G^* W_2{}^*)(W_2 G W_1)(I + W_2{}^* G^* W_1{}^* W_1 G W_2)^{-1}. \end{aligned} \tag{6.71}$$

By (6.70) we then have that

$$\overline{\sigma}(\tilde{M}_s) = \left(\frac{1}{1 + \underline{\sigma}^2(W_2 G W_1)} \right)^{1/2}$$

and by (6.71) we have

$$\overline{\sigma}(\tilde{N}_s) = \left(\frac{\overline{\sigma}^2(W_2 G W_1)}{1 + \overline{\sigma}^2(W_2 G W_1)} \right)^{1/2}.$$

■

The value of this Lemma is that the bounds on the closed-loop objectives in Theorem 6.12 can be rewritten in terms of γ, G, W_1, and W_2 only, and it can then be clearly seen how the loop shaping influences the closed-loop properties. This will be discussed in detail in the next section.

6.6 Discussions

6.6.1 The Role of the Shaping Functions in Design

As discussed earlier in this chapter, the shaping functions W_1 and W_2 can be used to trade-off between performance and robust stability requirements. In this section we will clearly demonstrate the relation between the shaping functions and performance and robust stability objectives. We now use Lemma 6.15 to re-evaluate the bounds of Theorem 6.12 over particular frequency ranges. We will separately consider frequency ranges of (i) high loop gain and (ii) low loop gain.

We firstly consider frequency regions of 'high loop gain' and assume that by 'high loop gain' we mean $\underline{\sigma}(G_S) = \underline{\sigma}(W_2 G W_1) \gg 1$, which will be typically at low frequency values. Then by (6.68) and (6.69) we have that $\overline{\sigma}(\tilde{M}_s) \leq (\underline{\sigma}(W_2 G W_1))^{-1}$, and

$\overline{\sigma}(\tilde{N}_s) \leq 1$. Recalculating the bounds in (6.52) - (6.55) using these inequalities, we obtain

$$\overline{\sigma}(K(I-GK)^{-1}) \leq \gamma c(W_1)/\underline{\sigma}(G) \tag{6.72}$$

$$\overline{\sigma}((I-GK)^{-1}) \leq \frac{\gamma}{\underline{\sigma}(G)\underline{\sigma}(W_1)\underline{\sigma}(W_2)} \tag{6.73}$$

$$\overline{\sigma}(K(I-GK)^{-1}G) \leq \gamma c(W_1) \tag{6.74}$$

$$\overline{\sigma}((I-GK)^{-1}G) \leq \frac{\gamma}{\underline{\sigma}(W_1)\underline{\sigma}(W_2)} \tag{6.75}$$

If we assume that W_1 and W_2 are selected such that $c(W_1)$ and $c(W_2)$ are of order 1, then the two performance measures, $\overline{\sigma}((I-GK)^{-1})$ and $\overline{\sigma}((I-GK)^{-1}G)$ will be improved by selecting $\underline{\sigma}(W_1)$ and $\underline{\sigma}(W_2)$ to be large, as was predicted earlier using loop shaping arguments. Further, note that in this *same* frequency region, robust stability measures $\overline{\sigma}(K(I-GK)^{-1})$ and $\overline{\sigma}(K(I-GK)^{-1}G)$ will be well behaved.

We now consider frequency regions of 'low loop gain', where by 'low loop gain' we mean $\overline{\sigma}(G_S) = \overline{\sigma}(W_2GW_1) \ll 1$, which will be typically at high frequency values. Then by (6.68) and (6.69) we have that $\overline{\sigma}(\tilde{M}_s) \leq 1$ and $\overline{\sigma}(\tilde{N}_s) \leq \overline{\sigma}(W_2GW_1)$. Rewriting the bounds in (6.52) - (6.55) we then obtain the following approximate relations

$$\overline{\sigma}(K(I-GK)^{-1}) \leq \gamma\overline{\sigma}(W_1)\overline{\sigma}(W_2) \tag{6.76}$$

$$\overline{\sigma}((I-GK)^{-1}) \leq \gamma c(W_2) \tag{6.77}$$

$$\overline{\sigma}(K(I-GK)^{-1}G) \leq \gamma\overline{\sigma}(G)\overline{\sigma}(W_1)\overline{\sigma}(W_2) \tag{6.78}$$

$$\overline{\sigma}((I-GK)^{-1}G) \leq \gamma\overline{\sigma}(G)c(W_2) \tag{6.79}$$

In this frequency range, robust stability measures $\overline{\sigma}(K(I-GK)^{-1})$ and $\overline{\sigma}(K(I-GK)^{-1}G)$ will be improved by selecting $\overline{\sigma}(W_1)$ and $\overline{\sigma}(W_2)$ to be small, again confirming the loop shaping philosophy. Also, note that $\overline{\sigma}((I-GK)^{-1})$ and $\overline{\sigma}((I-GK)^{-1}G)$ will be well behaved in this frequency range.

We conclude this section by comparing the results for the two frequency regions: At frequencies with high loop gain only performance is directly affected by the shaping functions, while at frequencies with low loop gain, only robust stability (as reflected by additive or multiplicative uncertainty measures) is directly affected. Note however that variations in W_1 and W_2 will change the frequency range over which the conditions $\underline{\sigma}(G_S) \gg 1$ and $\overline{\sigma}(G_S) \ll 1$ are valid, hence 'shifting' the frequency ranges over which performance and robust stability respectively, are of particular importance.

In summary, the results in this section confirm the Loop Shaping Design Procedure in 6.3 to be a valid design approach.

6.6.2 Model Reduction

A controller designed using the Loop Shaping Design Procedure will typically be of degree $n + 2n_w$, where n is the degree of the nominal plant, and n_w is the combined degrees of the shaping functions W_1 and W_2. Therefore, if the nominal plant has a high degree, then we expect a high degree controller, which may not be acceptable from an implementation point of view. However, the results from Chapter 5 on reduced order controller design using coprime factor model reduction can be *directly* applied to the Loop Shaping Design Procedure. The designer may obtain a reduced order model of K_∞ by either obtaining a lower order approximant of G_S, or by obtaining a reduced order approximant for K_∞ directly. In each case, the procedure to be used is identical to those in Propositions 5.5 and 5.6 respectively.

Note that by obtaining a lower order approximant of G_S we are essentially performing a *weighted* approximation of the nominal plant: Shaping functions W_1 and W_2 specify the frequency ranges where performance or robust stability characteristics are most important, and this is reflected in the approximation via the normalized coprime factors $\tilde{N}_s$ and $\tilde{M}_s$. (The effects of shaping on $\tilde{N}_s$ and $\tilde{M}_s$ are clearly illustrated in (6.68) and (6.69).)

6.7 Summary

In this chapter we have incorporated the normalized LCF robust stabilization problem into a loop shaping based design technique. This enables both performance and robust stability objectives to be traded-off, and preserves the exact solution associated with this particular H_∞ problem. The design method is straightforward and systematic, incorporating only the simpler aspects of 'classical' loop shaping. The normalized LCF robust stabilization problem has been shown to be particularly well suited for this approach because the H_∞ controller synthesized causes only a limited deterioration of the specified loop shape. Coprime factor model reduction techniques can be simply incorporated into this framework.

An interesting problem is the question of how the design indicator, ϵ_{max}, in the Loop Shaping Design Procedure, varies with changes in the selected shaping function. A promising approach seems to be to investigate the asymptotic behaviour of ϵ_{max} as either $\boldsymbol{W}_1$ or $\boldsymbol{W}_2$ become very large or small.

As a final comment, it should be noted that the Loop Shaping Design Procedure is restricted to a particular set of H_∞ objectives. However, we will shown in the next chapter that the design technique yields simple and effective controllers.

CHAPTER SEVEN

DESIGN EXAMPLES

7.1 Introduction

In this chapter we demonstrate the use of the Loop Shaping Design Procedure in three design examples. We will show that the procedure yields satisfactory feedback controllers in a relatively straightforward manner. In each design we follow the procedure outlined in Section 6.3, and incorporate model reduction as outlined in Section 6.6.2 when appropriate. In all designs presented in this chapter, the stabilizing H_∞ controller, obtained in Stage (2) of the design procedure, is chosen to be a sub-optimal controller for the normalized LCF robust stabilization problem. (See Section 4.3.) However, the (sub-optimal) stability margin, ϵ, in each case is chosen such that $0.95\epsilon_{max} \leq \epsilon < \epsilon_{max}$, and it clearly reflects the maximum achievable stability margin for the particular problem.

The first two examples are the results of a study performed for British Aerospace U.K., Space and Communications Division, between July, 1987, and February, 1988 which assessed the applicability of the Loop Shaping Design Procedure to the control of highly flexible space structures. (See McFarlane, Glover, and Noton, 1988.) In the first of these examples, the procedure is applied to a Single-Input, Single-Output (SISO) model, which enables a number of the features of the design procedure to be clearly illustrated. The second example, on a four-input, five-output model, demonstrates that the procedure can be readily applied to high order, non square, Multi-Input, Multi-Output (MIMO) systems.

The final design is a controller for a three-input, three-output linearized model of the vertical dynamics of an aircraft. This model has been recently studied by Maciejowski, (1989), and we will compare his controller designs with those obtained from the Loop Shaping Design Procedure.

7.2 Example 1: Attitude Control of a Flexible Spacecraft.

The purpose of this SISO example is to illustrate the mechanics of the Loop Shaping Design Procedure without the additional complications involved in MIMO design.

7.2.1 Problem Description

The system considered in this design is a satellite with two highly flexible solar arrays attached. The model for control analysis represents the transfer function from the torque applied to the roll axis of the satellite to the corresponding satellite roll angle. This model has been previously studied in Salehi, (1985) in assessing the application of observers to flexible spacecraft design. In order to keep the model simple, only a rigid body mode and a single flexible mode are included, resulting in a four state model. The state-space system is described by

$$\dot{x} = Ax + Bu + Bv$$

$$y = Cx$$

where u is the control torque (Nm), v is a constant disturbance torque (Nm), and y is the roll angle measurement (rad). The state-space matrices are given by

$$A = \begin{bmatrix} 0 & 1 & 0 & 0 \\ 0 & 0 & 0 & 0 \\ 0 & 0 & 0 & 1 \\ 0 & 0 & -\omega^2 & -2\zeta\omega \end{bmatrix}$$

$$B^T = [0 \quad 1.7319 \times 10^{-5} \quad 0 \quad 3.7859 \times 10^{-4}]$$

$$C = [1 \quad 0 \quad 1 \quad 0]$$

$$D = 0$$

where $\omega = 1.539\ rad/sec$ is the frequency of the flexure mode, and $\zeta = 0.003$ is the flexural damping ratio. The nominal open-loop poles are at

$$-0.0046 \pm 1.5390j$$

$$0, 0$$

and the finite zeros at

$$-0.0002 \pm 0.3219j.$$

Because of the highly flexible nature of this system, the use of control torque for attitude control can lead to excitation of the lightly damped flexural modes and hence loss of control. It is therefore desired to design a feedback controller which increases the system damping and maintains a specified pointing accuracy. That is, variations in the roll angle are to be limited in the face of torque disturbances. In addition, the stiffness of the structure is uncertain, and the natural frequency, ω, can only be approximately estimated. Hence it is desirable that the closed-loop be robustly stable to variations in this parameter.

7.2.2 Design Objectives

Based on Salehi, (1985), and Noton, (1987b), the objectives for this design were as follows:

(1) Performance - required pointing accuracy due to $0.3Nm$ step torque disturbances:

$$y < 0.04^o.$$

Additionally we require a response time < 1 minute.

(2) Robust stability - stable response for $\pm 10\%$ variations in natural frequency. A level of robust stability in the face of unstructured uncertainty is also desirable.

(3) Control level - control effort due to $0.3Nm$ step torque disturbances:

$$u < 0.5Nm.$$

In the design described next, it was attempted to achieve the best possible pointing accuracy, using the requirements in objectives (2) and (3) as constraints on the design.

7.2.3 The Loop Shaping Design Procedure

Two designs were attempted using the Loop Shaping Design Procedure and results from both are included here so that the effect of different shaping functions can be shown. Figures for this example are included at the end of the section.

DESIGN No. 1

1. Loop Shaping - inspection of the nominal open-loop magnitude (see Figure 7.1) indicates a cross-over frequency of $\approx 0.01 rad/sec$, implying a closed-loop response time of the order of 100 seconds - significantly slower than the desired response time. Hence some additional gain must be added to the loop shape to achieve an acceptably fast response. In addition, note that to achieve a pointing accuracy of 0.04^o $(= 0.0007 rad)$ in the face of $0.3 Nm$ torque input disturbances we require that the closed-loop transfer function $(\boldsymbol{I} - \boldsymbol{G}K)^{-1}\boldsymbol{G}$ satisfies the condition

$$|(\boldsymbol{I} - \boldsymbol{G}K)^{-1}\boldsymbol{G}| < \frac{0.0007}{0.3} = 0.0023 rad/Nm \tag{7.1}$$

at least at low frequencies. By (6.75) we have that (at low frequency):

$$|(\boldsymbol{I} - \boldsymbol{G}K)^{-1}\boldsymbol{G}| \leq \frac{\gamma}{|\boldsymbol{W}_1|.|\boldsymbol{W}_2|} = \frac{\gamma}{|\boldsymbol{W}|}$$

where shaping functions $\boldsymbol{W}_1$, $\boldsymbol{W}_2$, can be combined as a single shaping function, $\boldsymbol{W}$, without loss of generality for a SISO system. Hence by (7.1) we have the requirement

$$|\boldsymbol{W}| \geq 435\gamma$$

at low frequency. Anticipating that γ will satisfy $\gamma \lesssim 4$ (a reasonable estimate) we set

$$\boldsymbol{W} = 2000$$

as a first shaping function. Note that this will increase the open-loop cross-over frequency, therefore ensuring a faster response than if the nominal plant had been stabilized without shaping. The shaped plant is

$$\boldsymbol{G}_S = \boldsymbol{W}\boldsymbol{G}.$$

(Refer again to Figure 7.1 for the Bode plot of G_S.)

2. Robust Stabilization - A controller, K_∞, was obtained using the normalized LCF robust stabilization approach, with associated stability margin, ϵ, given by

$$\epsilon = 0.4274 \quad (\gamma = 2.34).$$

Note, by the results of Section 6.4 this indicates that there is $\approx 40\%$ allowable proportional uncertainty in $\tilde{N}_s$ and $\tilde{M}_s$ in the cross-over frequency range of the shaped plant, and that the loop shape achieved will differ from the desired loop shape by a maximum factor of $(\gamma^2 - 1)^{1/2} = 2.10$ at high and low frequencies.

3. The resultant feedback controller, K_1, was formed by

$$K_1 = K_\infty W$$

and the Bode plot of the stabilized loop-shape $GK_\infty W$ $(= GK_1)$ is given in Figure 7.1. State-space matrices for K_1 are given in Appendix D. Note from Figure 7.1 that there has been only minimal deterioration in the open-loop magnitude due to stabilization, while it has been necessary to make extensive alteration to the phase.

The required pointing accuracy was achieved in this design, although a 0.018^o steady-state error remained. Results are given in Section 7.2.4.

DESIGN No. 2

A second design was performed to improve steady-state disturbance rejection properties, and to attempt further improvements in pointing accuracy.

1. Loop Shaping - Noting by (6.75) that at low frequency, $|(I - GK)^{-1}G| < \gamma|W|^{-1}$, an additional term of the form

$$W_c = \frac{s + 0.4}{s}$$

was introduced into the shaping function, and as $|W_c(j0)| \rightarrow \infty$ implies $|(I - GK)^{-1}G(j0)| \rightarrow 0$, this *guarantees* full steady-state disturbance rejection. Further to improve the level of pointing accuracy, the shaping function gain was increased to 10000 and the combined shaping function was then

$$W = 10000 W_c.$$

A Bode Plot of the desired loop shape is shown in Figure 7.2. Comparing this with Figure 7.1 shows the effect of the additional integral action.

2. Robust Stabilization - A controller K_∞ was obtained and the associated stability margin ϵ was

$$\epsilon = 0.426 \quad (\gamma = 2.347)$$

indicating similar mid-frequency robust stability properties and similar limits to loop shaping deterioration to those shown in Design 1. Figure 7.3 shows a Bode Plot of K_∞. Note that $(\gamma^2 - 1)^{1/2} = 2.12$ for this problem and Figure 7.3 shows that at low frequency that $\underline{\sigma}(K_\infty) > 1/2.12$ and at high frequency that $\overline{\sigma}(K_\infty) < 2.12$. Hence, we have shown that the bounds on K_∞ given in Theorems 6.2 and 6.4 have been achieved exactly here.

3. The final controller, K_2, was formed by

$$K_2 = K_\infty W$$

and the state-space matrices of K_2 are given in Appendix D. Bode plots of the nominal, shaped and stabilized open-loop systems are given in Figure 7.2.

7.2.4 Results

The results from Design (2) are given below, and the results from Design (1) are given in parentheses.

(1) Robust Stability - The closed-loop was stable for variations in the natural frequency, ω, in the range $(60 - 118)\%$ $((95 - 150)\%)$ of its nominal value.

(2) Disturbance Rejection (Pointing Accuracy) - maximum deflection for $0.3Nm$ step input was 0.0026^o (0.019^o) and 0° (0.018^o) at steady-state.

(3) Control Effort - maximum control effort for $0.3Nm$ step input was $0.41Nm$ $(0.41Nm)$.

7.2.5 Analysis of Results

As anticipated the second design led to greatly improved pointing accuracy and total steady-state rejection of input torque disturbances at the output. See Figure 7.4 for a comparison of step responses for the two designs. Figure 7.5 shows that the corresponding control energy required to counter the disturbances was acceptable for both controllers. All time responses indicate that the closed-loop system was considerably damped. Minimum closed-loop damping ratio for Design (2) was 26%.

Figure 7.6 compares frequency domain measures for the two designs. In the second design it can be seen that performance measures $\overline{\sigma}((I-GK)^{-1})$ and $\overline{\sigma}((I-GK)^{-1}G)$ improved markedly at low frequency (Figure 7.6a,b) while high frequency robust stability measures, $1/(\overline{\sigma}(K(I-GK)^{-1}))$ and $1/(\overline{\sigma}(GK(I-GK)^{-1}))$, deteriorated slightly (Figure 7.6c,d). The effect of the integral action in Design (2) is clearly illustrated in Figure 7.6(b).

In Figure 7.7 the bounds on closed-loop objectives predicted in Theorem 6.12 are compared with the closed-loop objectives obtained in Design (2). Note that the performance bounds $\overline{\sigma}((I-GK)^{-1}) \leq \gamma\overline{\sigma}(\tilde{M}_s)$, $\overline{\sigma}((I-GK)^{-1}G) \leq \gamma\overline{\sigma}(\tilde{N}_s)(\underline{\sigma}(W))^{-1}$ are particularly accurate at low frequency while the robust stability bounds $1/(\overline{\sigma}(GK(I-GK)^{-1})) \geq (\gamma\overline{\sigma}(\tilde{N}_s))^{-1}$ and $1/(\overline{\sigma}(K(I-GK)^{-1})) \geq (\gamma\overline{\sigma}(\tilde{M}_s)\overline{\sigma}(W))^{-1}$, are particularly accurate at high frequency.

Comparison of the results from this design with the results in Salehi, (1985), where a conventional observer/state-feedback approach is used, showed that the H_∞ controller was better by a factor of 0.36:1 with respect to disturbance rejection, and more robust in terms of stability in the face of natural frequency variations. Salehi's study of this problem however, was more detailed, and also included an investigation of adaptive observers. The reader should therefore not attribute too much to this comparison, and consider this initial example to be merely an illustration of the Loop Shaping Design Procedure.

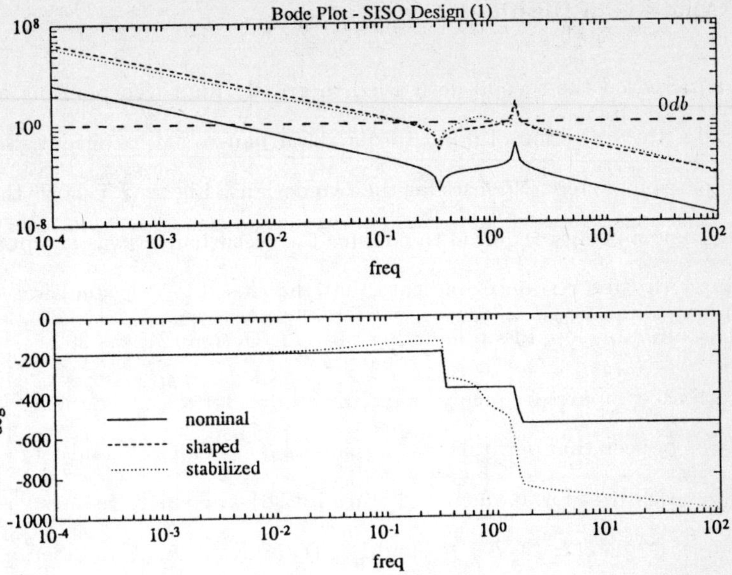

Figure 7.1 Bode Plots for Nominal, Shaped, Stabilized Systems - Design (1)

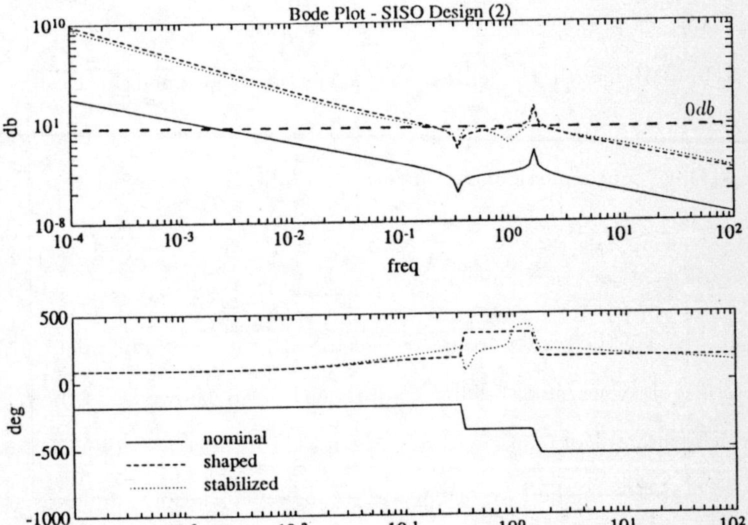

Figure 7.2 Bode Plots for Nominal, Shaped, Stabilized Systems - Design (2)

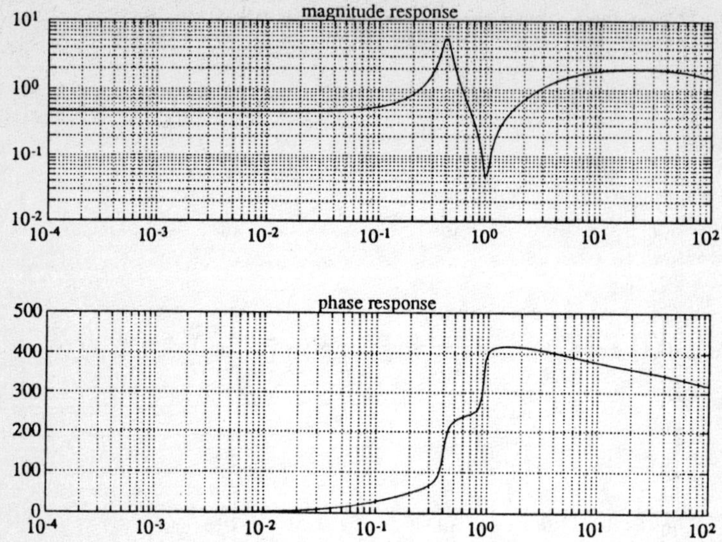

Figure 7.3 Bode Plot for K_∞ - Design (2)

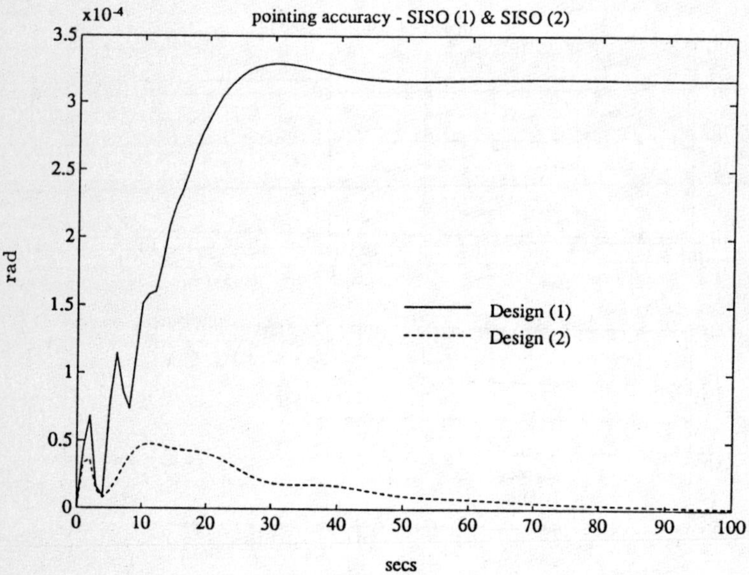

Figure 7.4 Output Response to Input Step - Designs (1), (2)

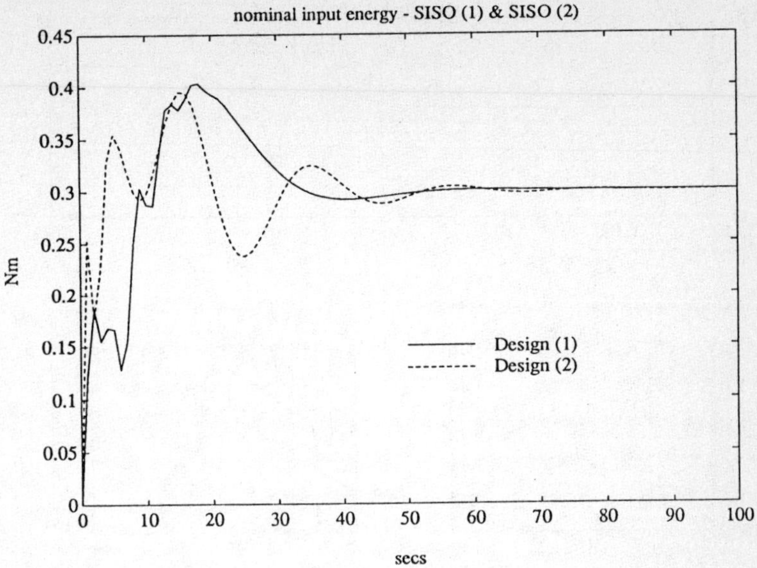

Figure 7.5 Input Response to Input Step - Designs (1), (2)

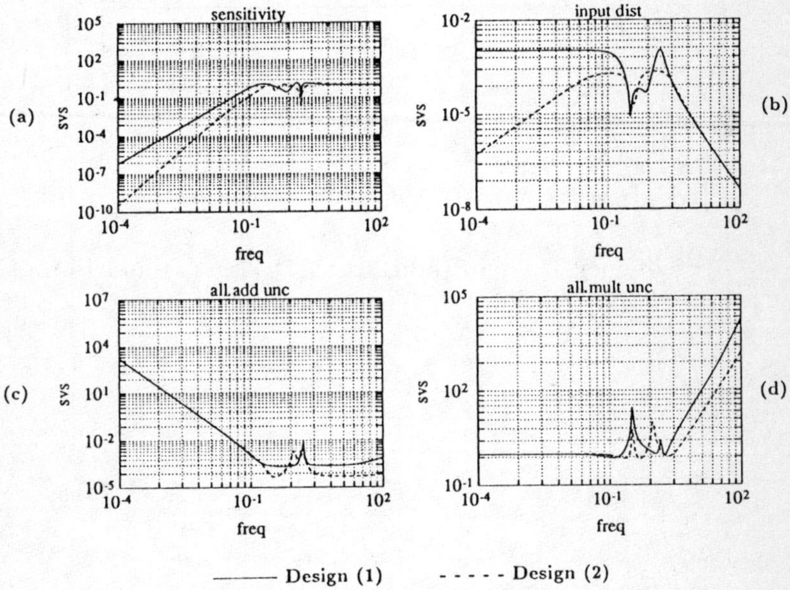

Figure 7.6 Frequency Domain Indicators - Designs (1), (2)

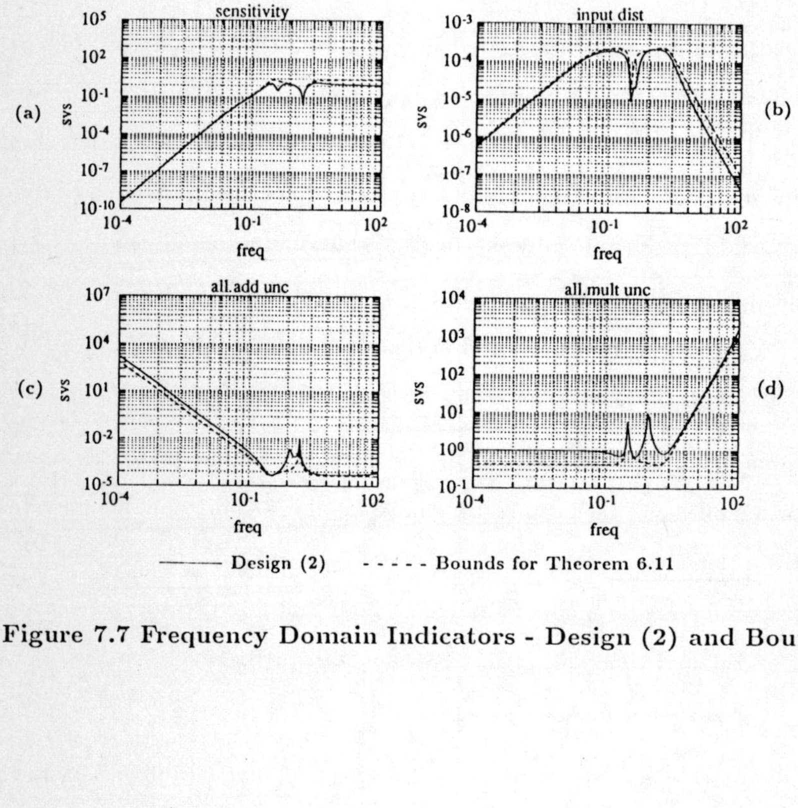

Figure 7.7 Frequency Domain Indicators - Design (2) and Bounds

7.3 Example 2: Attitude Control of a Flexible Space Platform

7.3.1 The Dynamic Model

The model being considered in this example is a simplified finite element model (MIMO) of a large space platform. The model was supplied by Dr. M. Noton, Space and Communications Division, British Aerospace P.L.C., and is intended to be representative of future large flexible spacecraft in general.

This elemental model is illustrated in Fig 7.8 as an assembly of point masses connected by massless struts exhibiting bending and torsional stiffnesses. The principal flexible effects arise from the solar arrays and the torsion of the main beam along the x-axis. This model has many modes but the studies have been restricted to rotations about the X and Z axes (roll and yaw), because the rotations about the Y axis (pitch) are decoupled and their control is simpler. The control variables shown in Fig 7.8 are:

u_1, u_2 - torques by reaction wheel in centrebody (limit $0.5Nm$)

u_3, u_4 - on-off thrusters at payloads P9 and P10 (limit $1N$)

The thrusters operate in an on-off mode but they have been treated as continuous linear actuators, because they can be time modulated in a short sample interval. Measurements are as below.

y_1, y_2 - roll and yaw angles at the centrebody (infra-red sensors)

y_3, y_4 - roll and yaw rates at the centrebody (rate gyros)

y_5 - roll rate at P9 (rate gyro)

Modal cost analysis (see Noton, 1987a) indicated that only five flexible modes need be considered and these tests were further restricted to the three most influential modes. The model has therefore 4 inputs, 5 outputs, and 10 states corresponding to 2 rigid body and 3 flexible modes. The nominal natural frequencies of the three flexible modes were

$$\omega_1 = 0.345\ rad/sec$$
$$\omega_2 = 0.531\ rad/sec$$
$$\omega_3 = 0.762\ rad/sec.$$

In practice, effective stiffnesses can be predicted only approximately beforehand, and the structures cannot be tested adequately on the ground. Therefore, in order to represent this uncertainty by a perturbed model the finite element analysis was repeated with a different principal stiffness, leading to significantly different modal frequencies, state and measurement matrices. Complete details of the nominal and perturbed state space systems are given in Appendix D, and the (non zero) open-loop singular values of the nominal and perturbed systems are shown in Figures 7.9 and 7.10. Damping of the modes does not come from the finite element analysis and is notoriously difficult to predict; it was set at not more than 1% for all modes.

7.3.2 Design Objectives

The two primary objectives of this design are that of robust stabilization and disturbance rejection. More specifically, because the actual stiffnesses of the platform are postulated to be very uncertain, it is desired to robustly stabilize the plant for a range of stiffnesses, and to ensure that the closed-loop damping is at an acceptable level. In addition, it is also desirable to minimize the effects of disturbances on outputs y_1 and y_2 (representing the angular positions of the platform) subject to constraints on peak inputs. The nominal specifications were therefore:

(1) Robust Stability - stabilize nominal and perturbed plants

- ensure damping $> 5\%$ always.

(2) Disturbance Rejection - for $0.1Nm$ step inputs on u_1, u_2, ensure maximum deflections satisfy:

$$y_1, y_2 \leq 0.01^o$$

(3) Input Magnitudes - for $0.1Nm$ step inputs on u_1, u_2, ensure the maximum control signals satisfy:

$$u_1, u_2 \leq 1.0Nm$$
$$u_3, u_4 \leq 0.5N$$

These are based on specifications set in current designs.

In the interests of good control the closed-loop bandwidth was permitted to embrace the flexible modes. This generally creates difficulties, but as is shown in this design, can lead to superior performance characteristics. Zero steady-state deflections due to disturbances were also desired.

7.3.3 The Loop Shaping Design Procedure

In total, five designs were produced for this problem, and all were satisfactory according to the specifications in Section 7.3.2, their bandwidth and steady-state properties being improved each time, while ensuring that robust stability and control energy requirements were satisfied. The final design is now described.

1. Loop Shaping - Inspection of the maximum singular value plot of the nominal plant (Figure 7.9) indicated that considerable additional gain was required to improve performance characteristics, especially the closed-loop bandwidth. Note that the nominal open-loop cross-over frequency at $\approx 0.005 rad/sec$ implies a very low closed-loop bandwidth. Gain of $\alpha = 4200$ in each output channel gave an open-loop gain cross-over frequency of $\approx 1 rad/sec$. (See Figure 7.11.) Simple PI compensators of the form

$$W_c(s) = \frac{s + 0.1}{s}$$

were also added to output channels y_1 and y_2 to ensure zero steady-state output due to disturbance in these channels. The motivation behind this selection is identical to that for the SISO design example. The shaped plant G_S was of the form

$$G_S = W_2 G$$

where

$$W_2 = diag\{4200 W_c; 4200 W_c; 4200 I_3\}.$$

The singular values of G_S are shown in Figure 7.11.

2. Robust Stabilization - A controller, K_∞, was generated and an optimal stability margin $\epsilon_{max} = 0.238$ was obtained, indicating, by Theorems 6.1 and 6.3, that deterioration in high and low frequency loop shapes is minimal and acceptable robust stability properties in the cross-over frequency region are expected. Following Procedure 5.5,

model reduction was performed on $[\tilde{N}_s, \tilde{M}_s]$, where $(\tilde{N}_s, \tilde{M}_s)$ is the normalized LCF of G_S, and two states were removed. The truncation error was 0.0189, only 8% of the final stability margin, ϵ, so that robust stability was not significantly degraded by this reduction. K_∞ was therefore a 10 state system.

3. Finally the H_∞ controller, K_∞, was cascaded with the shaping function W_2 to form the final controller, K_3, which had 12 states. See Appendix D for state-space matrices of K_3. Figure 7.12 illustrates the open-loop singular values of the stabilized system, K_3G.

7.3.4 Results

The results of the final design are presented here.

(1) Robust Stability - nominal and perturbed closed-loop systems were stable with minimum damping ratio of 12.6%-nominal and 8.6%- perturbed.

(2) Disturbance Rejection - maximum deflections for $0.1Nm$ step inputs on u_1 and u_2 were:

$$y_1(max) = 0.00075^o \text{ } (0^o \text{ steady-state})$$
$$y_2(max) = 0.00027^o \text{ } (0^o \text{ steady-state}).$$

Figure 7.13 shows the output responses for the nominal and perturbed systems.

(3) Control Effort- maximum magnitudes (for $0.1Nm$ step inputs) of u_1 and u_2 were:

$$u_1(max) = 0.055Nm$$
$$u_2(max) = 0.007Nm$$
$$u_3(max) = 0.020N$$
$$u_4(max) = 0.037N$$

These compare favorably with the specifications set out in Section 7.3.2. Figure 7.14 shows the control responses for nominal and perturbed systems due to $0.1Nm$ step input disturbances.

7.3.5 Analysis of Results

Figures 7.15-7.17 show closed-loop frequency domain measures relevant to this design. In Figure 7.15 the plot of $\overline{\sigma}((I - GK)^{-1}G)$ again highlights the effect of including integral action in the shaping function, giving zero steady-state transmission of input disturbance signals. The refinement of the disturbance rejection compared with the specified limits in this design was such that the major restriction in pointing accuracy is now the measurement system, rather than the feedback properties of the controller. See Noton, (1987a) for details. Figure 7.16 shows $1/(\overline{\sigma}(GK(I - GK)^{-1}))$ and indicates that at low frequency, output multiplicative uncertainties of 100% of the plant magnitude can be tolerated, and in the plant cross-over frequency region uncertainties of up to 30% can be tolerated. In Figure 7.17 the plot of $1/(\overline{\sigma}(K(I - GK)^{-1}))$ shows that at high frequency, additive uncertainty well in excess of the plant magnitude can be tolerated. (It is common to interpret robust stability in terms of a combination of these two uncertainty descriptions.)

Finally, it is interesting to compare the loop shapes obtained at different stages of the design: Comparing Figures 7.11 and 7.12 it can be seen that the robust stabilization stage has not significantly altered the desired loop shape. In particular note that the oscillatory plant modes are still present in the stabilized open-loop system. Cancellation of these modes by the controller can lead to poor robust stability and robust performance properties.

7.3.6 Spillover Mode Analysis

The controller design in Section 7.3.3 was based on a model containing only three flexible modes. However, it was stated earlier that the first *five* flexible modes were significant in the behaviour of the system. Hence, for evaluation purposes, extended nominal and perturbed models (including all five significant modes) were assembled, and the controller was assessed with these new models. The natural frequencies of these additional modes were nominally

$$\omega_4 = 1.635\ rad/sec$$
$$\omega_5 = 2.033\ rad/sec.$$

The controller stabilized both the nominal and perturbed evaluation models, and time responses for disturbance rejection and corresponding control effort are given in Figures 7.18 and 7.19 respectively. Not unexpectedly, the closed-loop damping in each case was low - the controller was not designed to dampen these spillover modes.

7.3.7 Design on a Model with Reduced Number of Inputs

Further investigations have considered the highly desirable possibility of controlling the space platform only by means of the reaction wheels in the centrebody (inputs u_1, u_2 only). Two further designs have been performed on a reduced input model, and results indicate that such a control scheme would be possible. A controller, K_4, (see Appendix D for state-space description) was first produced using the same shaping function used in the previous design and it was able to stabilize both the nominal and perturbed systems. Figure 7.20(a) shows that for the nominal system, the maximum pointing error was increased by 240% compared with the original four input design, and in Figure 7.20(b) the control effort is also shown to be slightly higher for the two reaction wheels. These results are still highly acceptable compared with the initial specifications, but subsequent spillover analysis showed that the controller was unable to stabilize the higher order, perturbed system, indicating poor robust stability at high frequency.

Hence, in order to improve high frequency robust stability properties (see (6.76)) for this design, the shaping function was reduced to

$$W = diag\{2100W_c, 2100W_c, 2100I_3\}$$

where $W_c = \frac{s+0.1}{s}$ and a further controller K_5 was produced. (See Appendix D for the state-space matrices of K_5.) This new controller was able to stabilize the perturbed spillover system, with the trade-off for the improved robust stability being a further loss in pointing accuracy - the pointing error was 460% greater than that in the original four input design. (See Figure 7.21a.) Control effort requirements are illustrated in Figure 7.21(b).

To conclude, it is certainly possible to restrict the control action to inputs u_1 and u_2 alone, with the trade-off being losses in robust stability and/or pointing accuracy.

7.3.8 Summary

The controller design for the flexible space-platform has proven to be relatively simple to perform, and the results in Section 7.3.4 are favorable. Although the evaluation of the controllers was limited to linear simulations only, the results indicate that this Loop Shaping Design Procedure can be successfully applied to flexible spacecraft control.

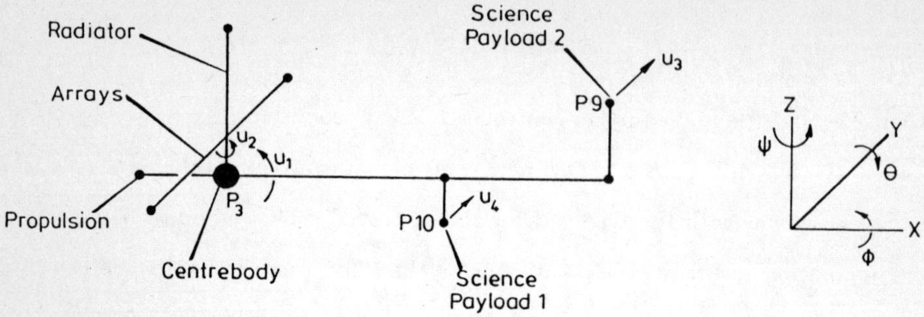

Figure 7.8 Elementary Model of a Spacecraft

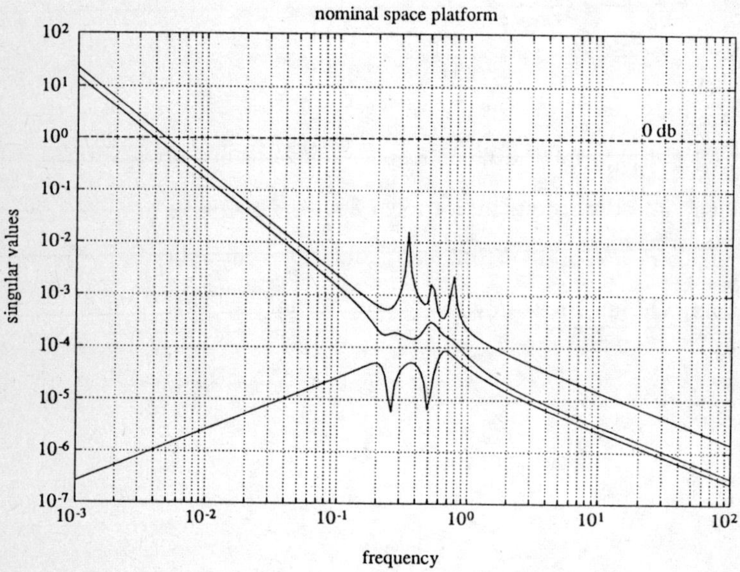

Figure 7.9 Open-Loop Singular Values - Nominal Plant

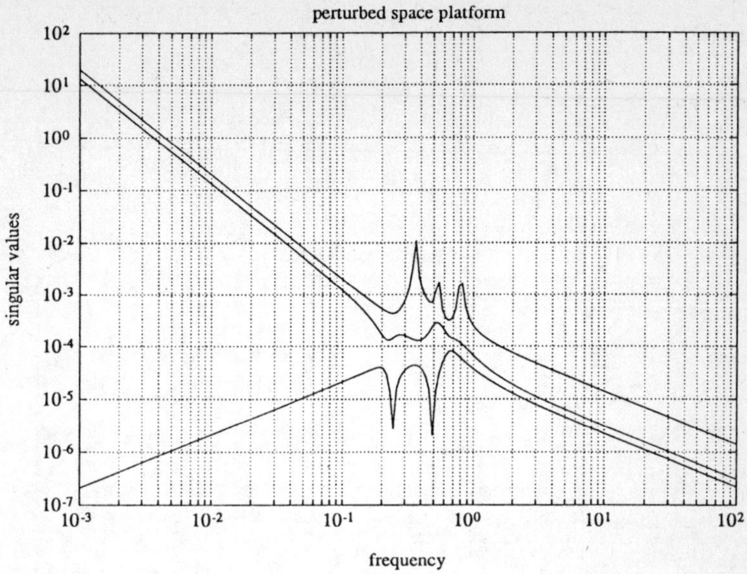

Figure 7.10 Open-Loop Singular Values - Perturbed Plant

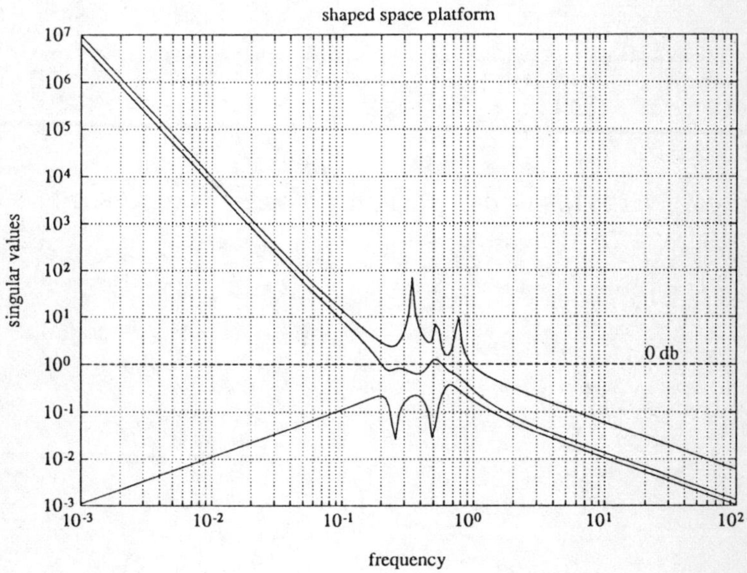

Figure 7.11 Open-Loop Singular Values - Shaped Plant

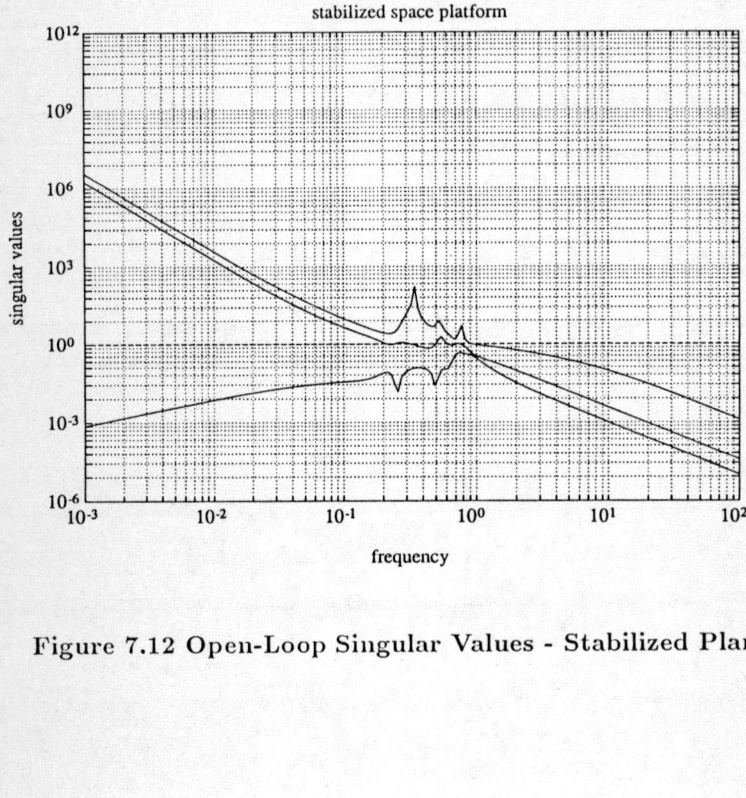

Figure 7.12 Open-Loop Singular Values - Stabilized Plant

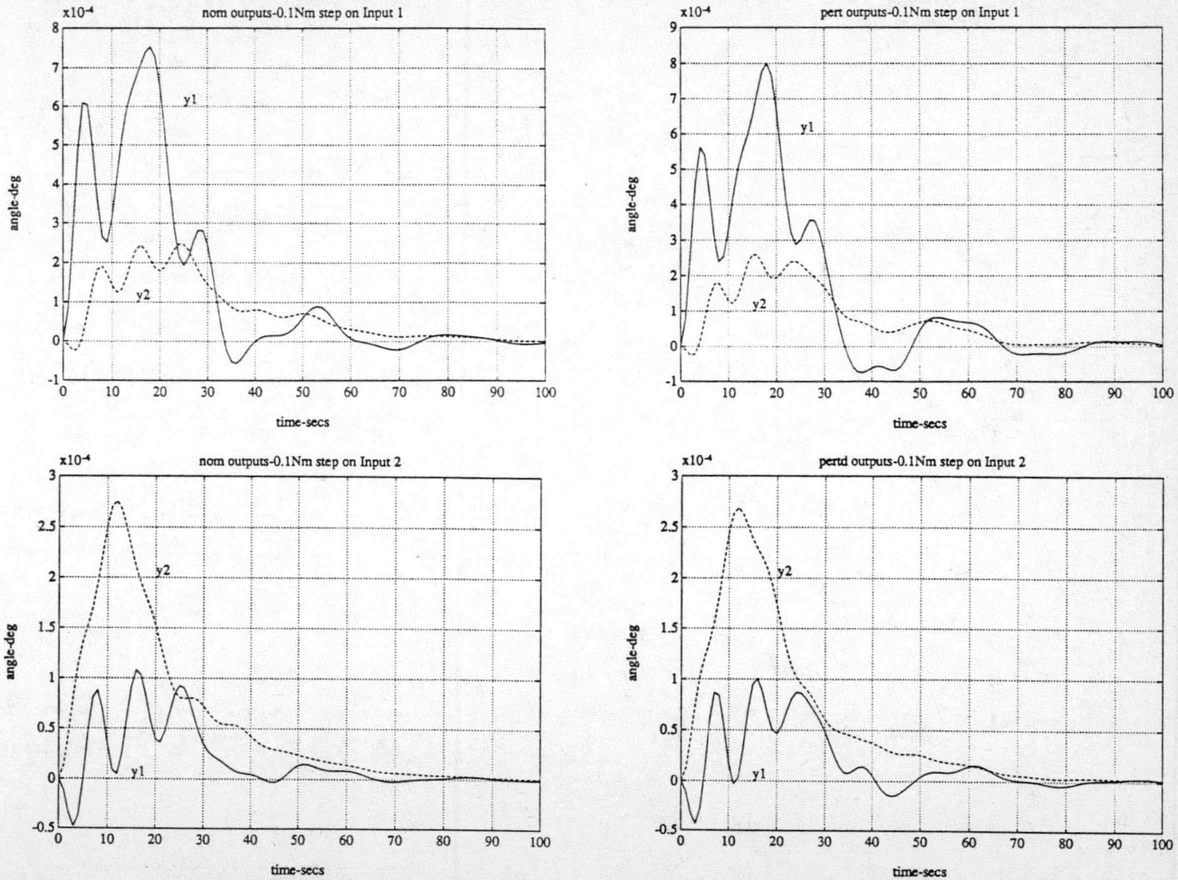

Figure 7.13 Output Responses to 0.1Nm Input Steps (Nominal and Perturbed Systems)

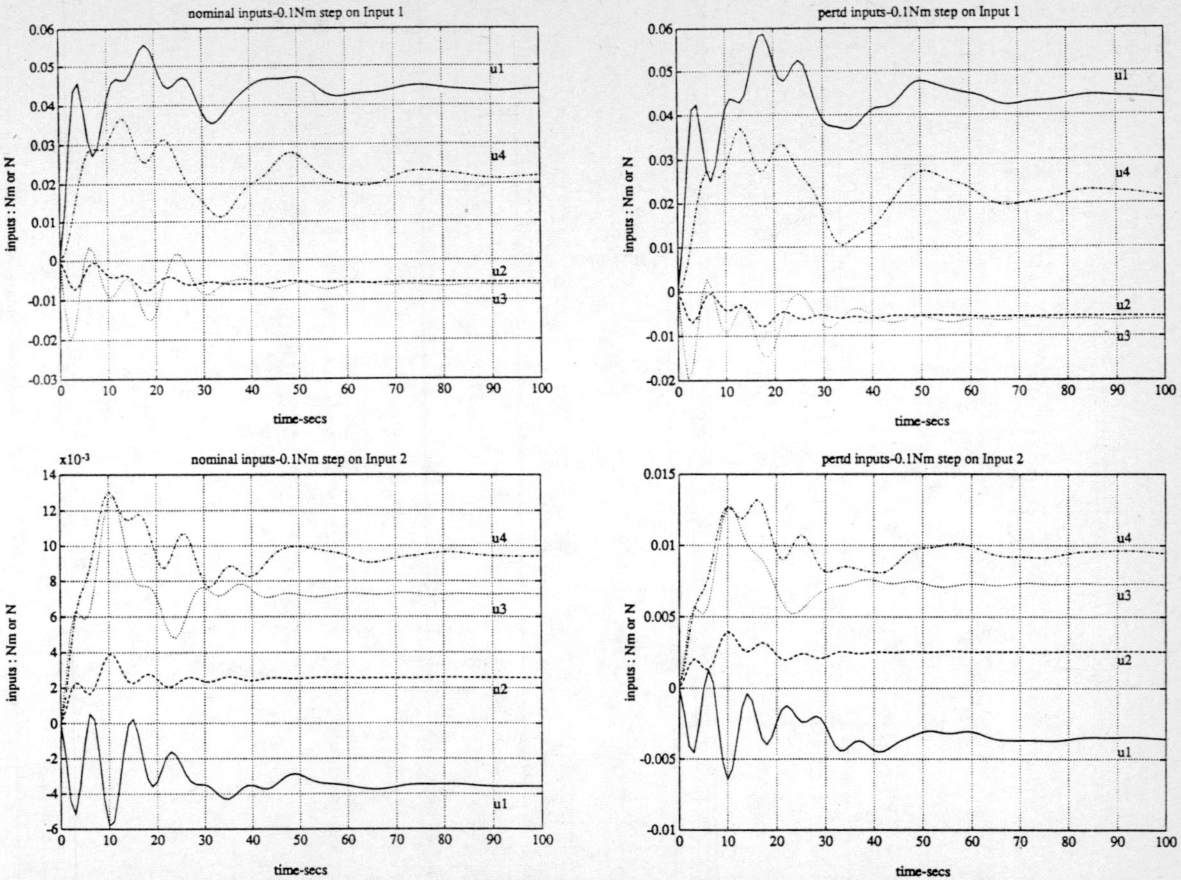

Figure 7.14 Input Responses to 0.1Nm Input Steps (Nominal and Perturbed Systems)

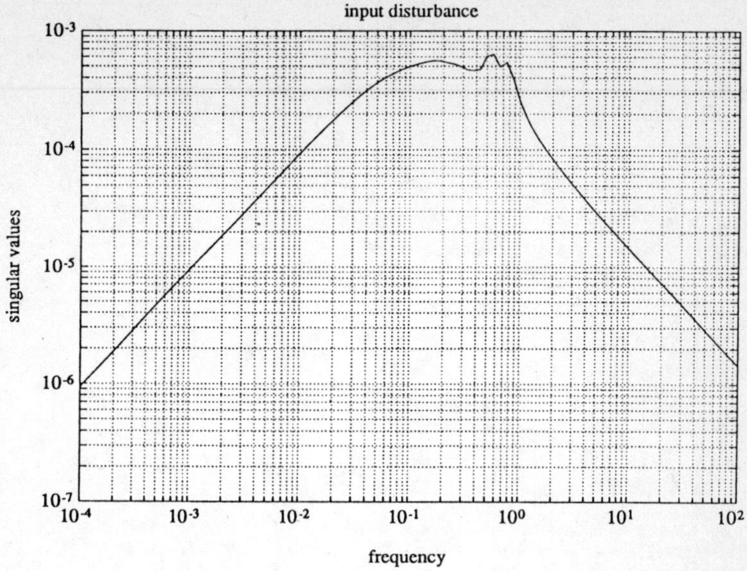

Figure 7.15 $\overline{\sigma}((I-GK)^{-1}G)$ **versus Frequency**

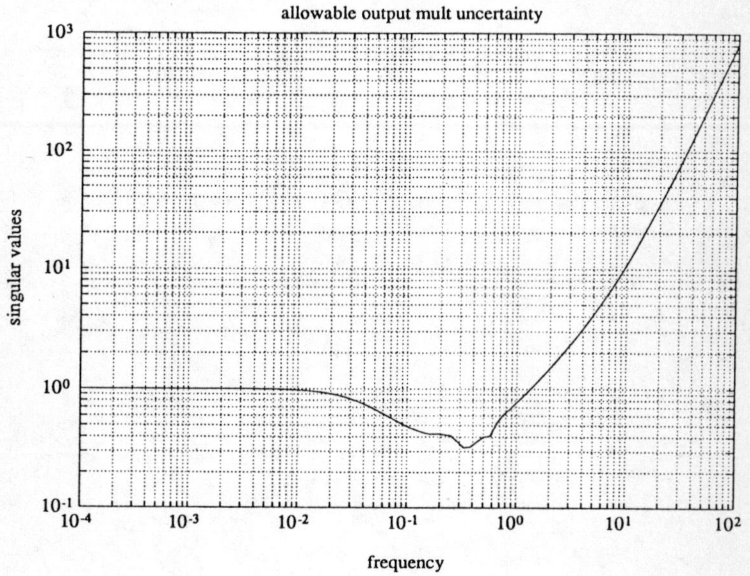

Figure 7.16 $1/(\overline{\sigma}(GK(I-GK)^{-1}))$ **versus Frequency**

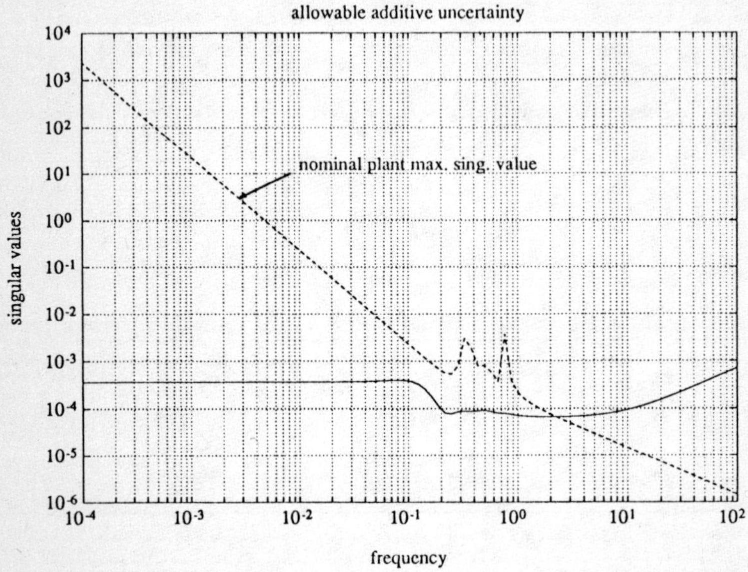

Figure 7.17 $1/(\overline{\sigma}(K(I-GK)^{-1}))$ **and** $\overline{\sigma}(G)$

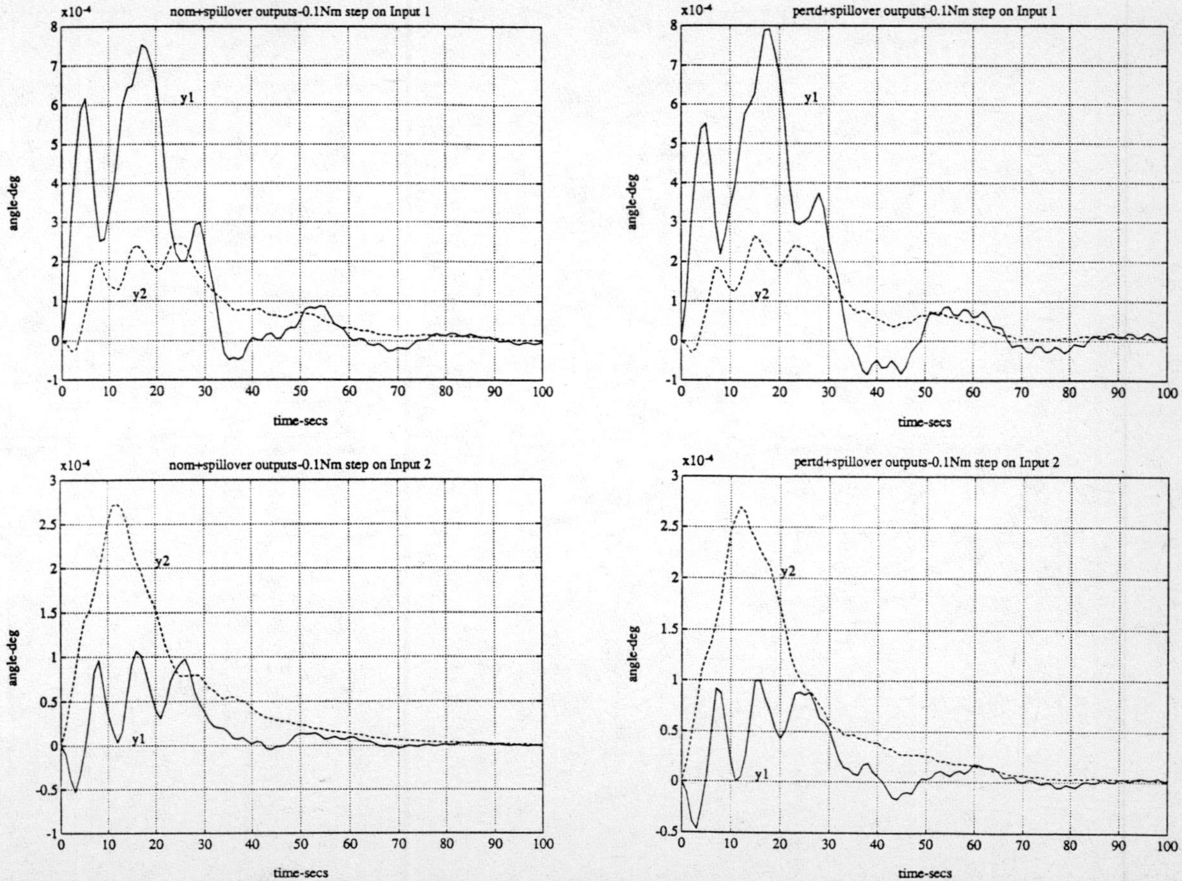

Figure 7.18 Output Responses to 0.1Nm Input Steps (Nominal and Perturbed Spillover Systems)

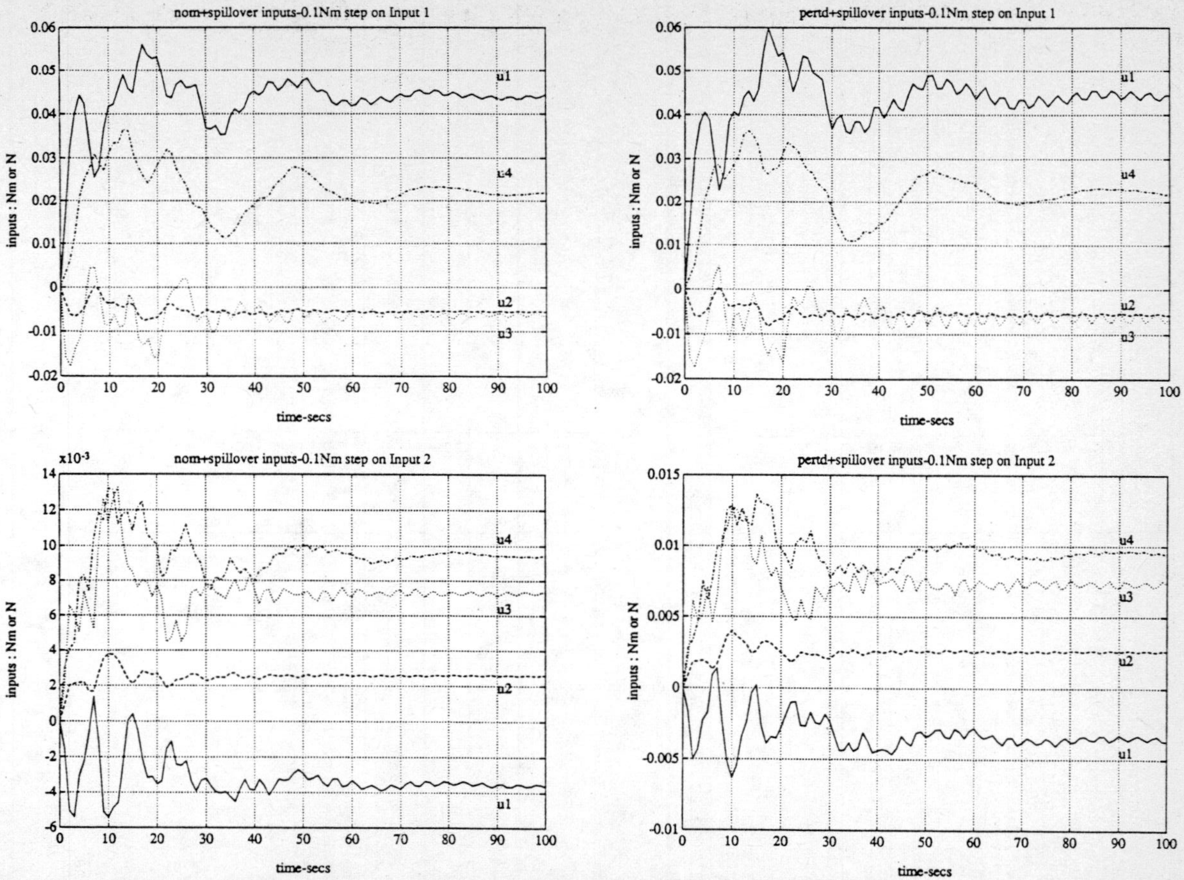

Figure 7.19 Input Responses to 0.1Nm Input Steps (Nominal and Perturbed Spillover Systems)

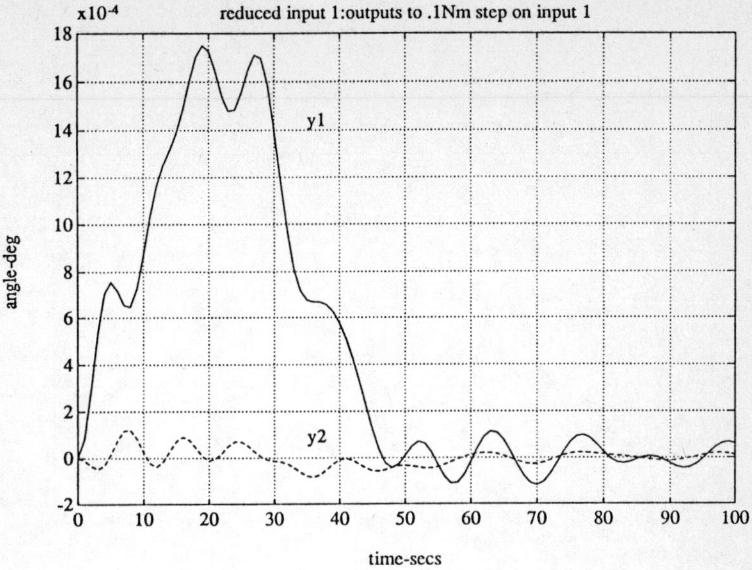

(a) Output Response to Step Input

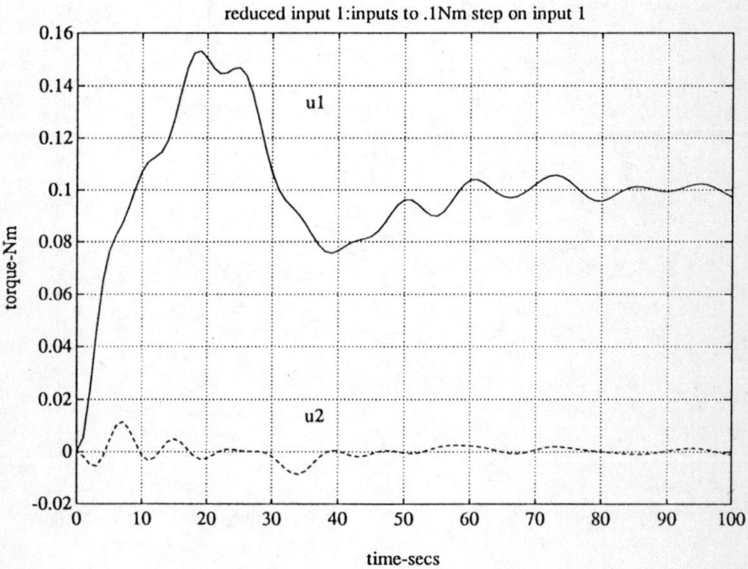

(b) Input Response to Step Input

Figure 7.20 Reduced Inputs Design (1)

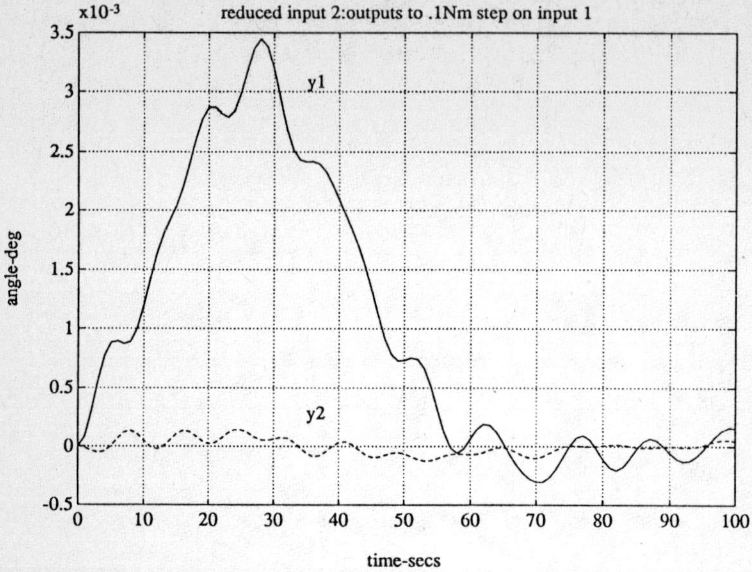

(a) Output Response to Step Input

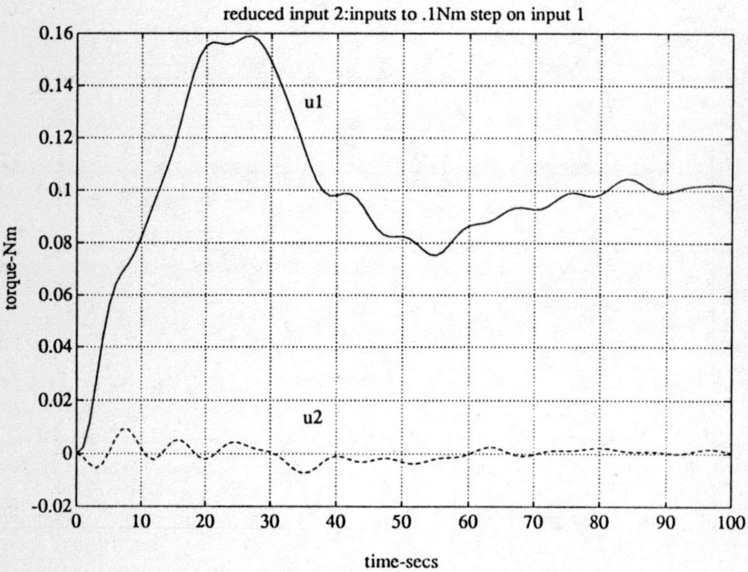

(b) Input Response to Step Input

Figure 7.21 Reduced Inputs Design (2)

7.4 Example 3: Vertical Plane Dynamics Control of an Aircraft

In this example we will consider a MIMO model of the vertical dynamics of an aircraft. This problem has previously been considered by Hung and MacFarlane, (1982) and Maciejowski, (1989). In Section 7.4.6 we will compare LSDP designed controllers with those obtained by Maciejowski using Loop Transfer Recovery (LQG) and the H_∞ Mixed Sensitivity approach.

7.4.1 Problem Description

The vertical plane dynamics model has 3 inputs, 3 outputs and 5 states. The inputs are

u_1 - spoiler angle ($^o \times 10^{-1}$)

u_2 - forward acceleration (m/sec^2)

u_3 - elevator angle (^{o})

and the states are

x_1 - relative altitude (m)

x_2 - forward speed (m/sec)

x_3 - pitch angle (^{o})

x_4 - pitch rate ($^o/sec$)

x_5 - vertical speed (m/sec)

Finally, the three outputs, y_1 - y_3, are simply measurements of the first three states. A state-space model for the system is given in Appendix D, and we note that the system has open-loop poles at

$$\begin{array}{l} -0.78 \pm 1.03j \\ -0.0176 \pm 0.1826j \\ 0 \end{array}$$

and no finite transmission zeros.

7.4.2 Design Objectives

The specifications for the LSDP design were as follows:

(1) Performance - Loop bandwidth of approximately 10 rad/sec for each loop and zero steady-state error in the face of step demands or disturbances. It is desirable to have little interaction between outputs and good damping of responses.

(2) Robust Stability - Minimum allowable output multiplicative uncertainty: 50% at all frequencies.

(3) Control Effort - The following limits were placed on the control variables in the face of unit step demands:

$$u_1 < 40^o$$
$$u_2 < 10\ m/sec$$
$$u_3 < 40^o.$$

7.4.3 The Loop Shaping Design Procedure

Two designs are described here, the first using non-dynamic shaping, the second using additional dynamic shaping.

DESIGN 1.

1. Loop Shaping - Figure 7.22 shows the open-loop singular values of the nominal plant. To improve low frequency performance and increase the closed-loop bandwidth a shaping function

$$W = diag\{24, 12, 24\}$$

was used. (The second output was scaled with a lower gain in order to decrease its bandwidth relative to the other outputs as an initial design indicated that y_2 responds significantly faster than the other two outputs.) By (6.73) and (6.75) higher gains would be expected to give further performance improvements, but by (6.76) it can be seen that this would lead to increased control effort and decreased robustness. Figure 7.23 shows the open-loop singular values of the shaped aircraft.

2. Robust Stabilization - A fifth order controller K_∞ was obtained. (No model reduction was possible without significant error.) A stability margin

$$\epsilon = 0.404 \quad (\gamma = 2.476)$$

was obtained indicating approximately 40% allowable proportional uncertainty in $\tilde{N}_s$ and $\tilde{M}_s$ in the cross-over frequency range, and minimal loop shape distortion at high and low frequencies. The loop shape of the stabilized aircraft is given in Figure 7.24.

3. The final controller $K_6 = K_\infty W$ has five states. (See Appendix D for the state-space matrices.) Inspection of the frequency domain indicators in Figure 7.25 at low frequency shows that a steady state error would exist in the face of both output disturbances ($\overline{\sigma}((I - GK)^{-1})$ in Figure 7.25a) and input disturbances ($\overline{\sigma}((I - GK)^{-1}G)$ in Figure 7.25b). In Figure 7.25(d), the plot of $1/(\overline{\sigma}(GK(I - GK)^{-1}))$ shows that multiplicative uncertainty requirements have been met, and noting that Figure 7.25(c) gives $1/(\overline{\sigma}(K(I - GK)^{-1}))$ against frequency, we have that

$$\overline{\sigma}(K(I - GK)^{-1}) \leq 50$$

at all frequencies, indicating that controller action due to output disturbances (or reference signals) is limited.

In order to improve the low frequency performance, a second design was performed:

DESIGN 2.

1. Loop Shaping - The shaping function of Design 1 was appended with the dynamic term,

$$W_c = \frac{s + 0.4}{s}$$

in each channel, yielding a shaping function

$$W = diag\{24W_c, 12W_c, 24W_c\}.$$

The integral action was expected to improve low frequency performance, and in particular would be expected to result in zero steady-state disturbance responses. The singular values of the shaped system are illustrated in Figure 7.26.

2. Robust Stabilization - A slightly lower ϵ value,

$$\epsilon = 0.378 \quad (\gamma = 2.64)$$

was obtained for this second design. Model reduction was performed at two stages during the design of K_∞: A single state was truncated from $[\tilde{N}_s, \tilde{M}_s]$, where $(\tilde{N}_s, \tilde{M}_s)$ is the normalized LCF of G_S, and a further state was truncated from $\begin{bmatrix} U \\ V \end{bmatrix}$, where (U, V) is the RCF of K_∞, and $\begin{bmatrix} U \\ V \end{bmatrix}$ is a sub optimal Nehari extension of $\begin{bmatrix} -\tilde{N}_s^* \\ \tilde{M}_s^* \end{bmatrix}$. (See Section 5.3.) Balanced truncation was used in each case. The resulting H_∞ controller K_∞ had six states.

3. The final controller, $K_7 = K_\infty W$, had nine states (see Appendix D for details) and the singular values of the stabilized open-loop system are given in Figure 7.27.

7.4.4 Results

The results of the designs are briefly summarized below. Results for Design (1) are in parentheses.

(1) Performance

(a) Bandwidth:loops $2, 3 \simeq 10\ rad/sec$ $(10\ rad/sec)$

$1 \simeq 3\ rad/s$ $(3\ rad/sec)$

(b) Interaction $\simeq 25\%$ (20%)

(c) Overshoot $\simeq 25\%$ (25%)

(d) Steady State Error $= 0\%$ (2%).

(2) Robust Stability - Minimum allowable output multiplicative uncertainty: 70% (70%).

(3) Control Effort- maximum magnitudes (for unit step inputs) on u_1, u_2 and u_3 were:

$$u_1(max) = 2.3^o\ (2.3^o)$$

$$u_2(max) = 5.2\ m/sec^2\ (5.2\ m/sec^2)$$

$$u_3(max) = 38.5^o\ (37.5^o).$$

7.4.5 Analysis of Results

Comparing the frequency response plots of performance measures $\overline{\sigma}((\boldsymbol{I}-\boldsymbol{GK})^{-1})$ and $\overline{\sigma}((\boldsymbol{I}-\boldsymbol{GK})^{-1}\boldsymbol{G})$ for Design (2) in Figures 7.28(a) and (b), with those for Design (1) in Figures 7.25(a) and (b), it can be seen that low frequency performance has been improved in the second design. Measures for allowable additive and multiplicative uncertainty, $1/(\overline{\sigma}(\boldsymbol{K}(\boldsymbol{I}-\boldsymbol{GK})^{-1}))$ and $1/(\overline{\sigma}(\boldsymbol{GK}(\boldsymbol{I}-\boldsymbol{GK})^{-1}))$, in Figures 7.28(c) and (d) respectively, remain virtually unchanged from the previous design. In addition, note in Figures 7.28(e) and (f), that closed-loop properties at plant input, $1/\overline{\sigma}(\boldsymbol{K}(\boldsymbol{I}-\boldsymbol{GK})^{-1}\boldsymbol{G})$ and $\overline{\sigma}((\boldsymbol{I}-\boldsymbol{KG})^{-1})$, are also satisfactory.

Figure 7.29 shows output responses to unit step output disturbances for Designs (1) and (2). Note that the additional integral action in Design (2) ensures a zero steady-state error, but that otherwise, there is little difference between the responses. The similar behaviour in each case can be attributed to the integral action already present in the nominal plant. (If, however, we were to compare the effects of *input* disturbances on the plant output, then the integral action in the controller of Design (2) would ensure greatly superior performance.) Interaction and overshoot are of the order 20 – 25% which is acceptable based on the results in the designs Maciejowski, (1989). The response of the altitude output in both Design (1) and (2) is slower than specified, and this can be attributed to using only the simplest shaping in the mid frequency region rather than trying to compensate individual loops.

Figure 7.30 gives the corresponding control requirements for the output step disturbances. Inputs u_1 and u_2 are well within specified limits in each case, while u_3, the elevator angle, is right up against the specified limit (compare with Section 7.4.2). This limit is the reason for accepting slightly slower responses than were specified: Note that the transfer function from output disturbances to the controls is given by $\boldsymbol{K}(\boldsymbol{I}-\boldsymbol{GK})^{-1}$, and, by (6.76), at high frequency, we have

$$\overline{\sigma}(\boldsymbol{K}(\boldsymbol{I}-\boldsymbol{GK})^{-1}) < \gamma\overline{\sigma}(\boldsymbol{W}).$$

This tells us that we could not expect to achieve any further increases in closed-loop bandwidth (requiring increased W) without violating the elevator angle limit.

7.4.6 Comparison with Maciejowski's Results

We will now compare the controllers designed using the Loop Shaping Design Procedure (termed LSDP(1) and LSDP(2)) with two controllers designed by Maciejowski, (1989), on the same aircraft vertical dynamics model.

The first of these controllers was designed using the LQG-based Loop Transfer Recovery (LTR) procedure. In this method the cost weights and noise covariances of an LQG problem are manipulated to achieve a desired loop shape. As was shown in Section 4.6, LQG problems are equivalent to the frequency domain H_2 optimization problem. In particular, Stein and Athans, (1987), have shown that, under certain asymptotic conditions, the LTR problem is approximately equivalent to finding

$$\inf_{K \text{ stabilizing}} \left\| \begin{bmatrix} (I-GK)^{-1}W & -GK(I-GK)^{-1} \\ 0 & 0 \end{bmatrix} \right\|_2 \tag{7.2}$$

where W reflects a desired loop shape that is specified from within the LQG framework. This problem is, of course, the H_2 version of the H_∞ Weighted Mixed Sensitivity optimization problem which was introduced in Section 4.5. It is convenient (and permissible) in this analysis to interpret Maciejowski's LTR problem as the H_2 optimization problem of (7.2). The singular values of the selected weighting function W for Maciejowski's LTR design are shown in Figure 7.31.

The second of Maciejowski's designs solves the 'regular' H_∞ Mixed Sensitivity optimization problem of finding

$$\inf_{K \text{ stabilizing}} \left\| \begin{bmatrix} W_1(I-GK)^{-1} \\ W_2GK(I-GK)^{-1} \end{bmatrix} \right\|_\infty \tag{7.3}$$

where

$$W_1 = W_{1c} \times I_3$$

$$W_2 = W_{2c} \times I_3$$

and

$$W_{1c} = \frac{(s+6)^2}{(s+0.00006)(s+0.6)}$$

$$W_{2c} = \frac{2000(s+10)(s+50)}{(s+1000)^2}.$$

The singular values of W_1 and W_2 are given in Figure 7.32. Note that the effect of W on the cost function in (7.2) is very similar to the combined effects of W_1 and W_2 on the cost function in (7.3): The intention in each case is to minimize sensitivity (S) at frequencies below $10rad/s$ and minimize complementary sensitivity (T) at frequencies above $10rad/s$. The minimization is with respect to the H_2 norm in the LTR case, and with respect to the H_∞ norm in the H_∞ Mixed Sensitivity case. However, despite the similarities between the effects of W_1 and W_2, and those of W, it is important to note that the LTR weight, W, is implicitly related to the nominal plant by the nature of the LTR design procedure (see Maciejowski, 1989), while W_1 and W_2 are simply chosen to reflect design requirements. This difference will be highlighted in the next section.

In each of these designs, the specifications were the same as the performance specifications in Section 7.4.2. (The nature of the objectives to be minimized in (7.2) and (7.3) is such that control effort objectives cannot be taken into consideration in either the LTR or H_∞ Mixed Sensitivity problems.)

The LTR controller designed by Maciejowski had 8 states, while the H_∞ Mixed Sensitivity (S&T) controller had 12 states.

We will now compare the closed-loop behaviour of the LTR and H_∞ Mixed Sensitivity controllers to those designed using the Loop Shaping Design Procedure. Figures 7.33 and 7.34 show plots of sensitivity, $\overline{\sigma}((I - GK)^{-1})$, and inverse complementary sensitivity, $1/(\overline{\sigma}(GK(I - GK)^{-1}))$, for each of the four controllers. As would be expected, the LTR and S&T controllers perform well here. (Recall that in (7.2) and (7.3) we have minimized these objectives.) Figure 7.35 gives the output time responses due to output step disturbances for the LTR and S&T controllers. Comparing these with the time responses for LSDP (1) and (2) in Figure 7.29, these responses confirm the superior performance properties over the LSDP designs.

Figure 7.36 shows $1/(\overline{\sigma}(K(I - GK)^{-1}))$ for each of the four designs. The two LSDP controllers are much more robust to the presence of high frequency additive uncertainties than the other two. Further, noting that $\overline{\sigma}(K(I - GK)^{-1})$ represents the maximum energy gain from output disturbances to the controls, we anticipate that significantly more control action would be required for the LTR and S&T controllers to reject output disturbances than for the LSDP controllers. This is confirmed in Figure 7.37 where requirements on elevator angle for the LTR and S&T controllers become extremely large indeed.

This last result highlights a limitation of both LTR and H_∞ Mixed Sensitivity approaches, in that it is not possible to include control energy considerations in the cost functions in either case.

A further defect of the H_∞ Mixed Sensitivity approach is that, for simple weight selections, lightly damped plant modes will be *cancelled* with zeros in the controller, rather than using the effects of feedback to *dampen* these modes in the closed-loop. (This cancellation is clearly illustrated by the singular values of the stabilized open-loop for the S&T problem given in Figure 7.38. *No* resonance effects are evident.) As discussed in Maciejowski, (1989), this is 'due to closed-loop specification being imposed without regard for the plant characteristics'. Note that in the LTR design, the function W includes the resonant modes of the nominal plant (see Figure 7.31), and the resulting stabilized open-loop system also (see Figure 7.39) includes the plant resonant modes. The ability of both the LSDP and LTR methods to to 'account for plant characteristics' is a subject worthy of further investigation.

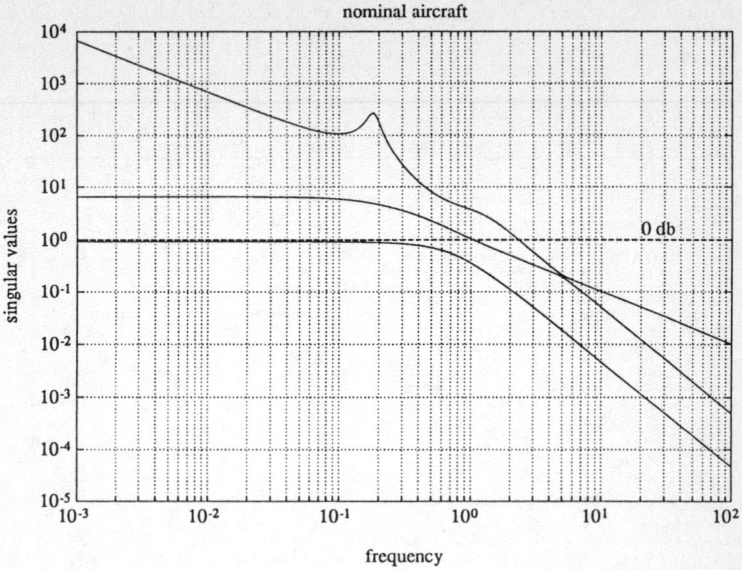

Figure 7.22 Open-Loop Singular Values–Nominal Plant

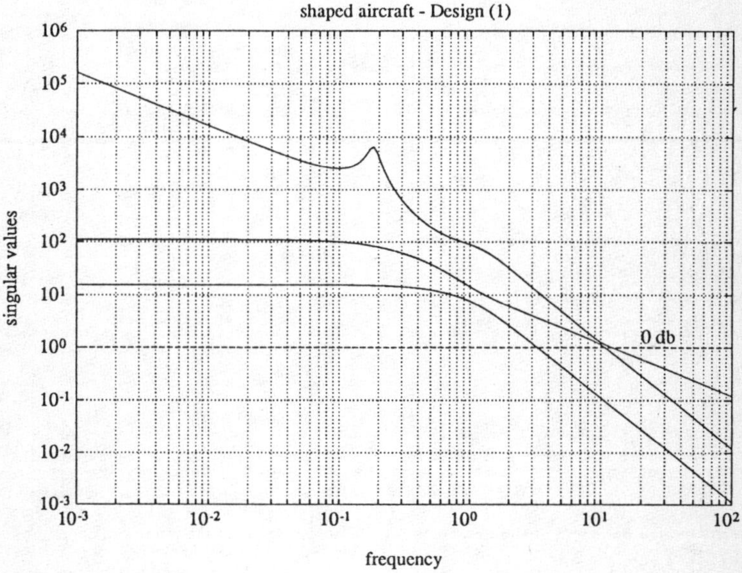

Figure 7.23 Open-Loop Singular Values–Shaped Plant (1)

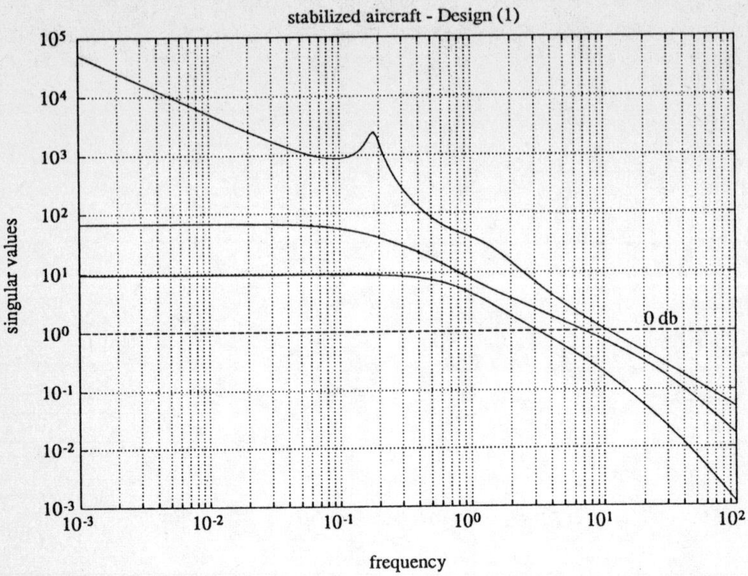

Figure 7.24 Open-Loop Singular Values–Stabilized Plant (1)

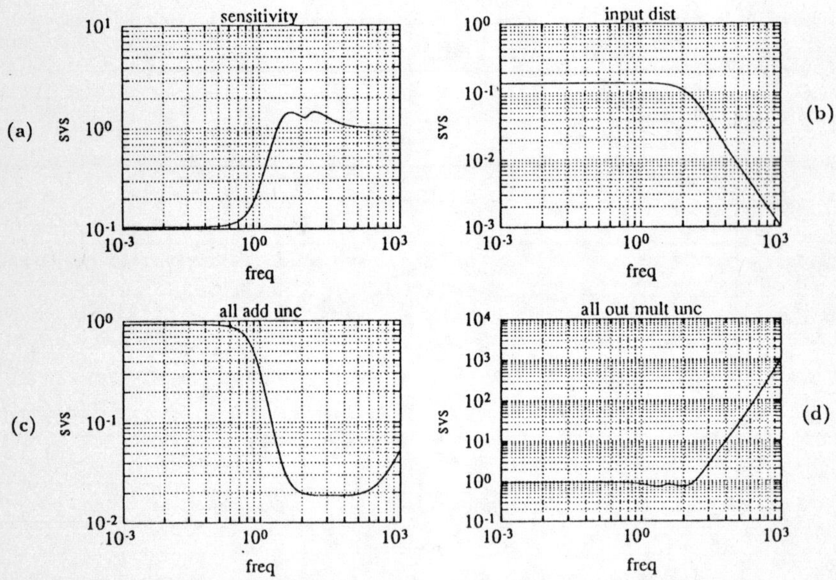

Figure 7.25 Frequency Domain Indicators–Design (1)

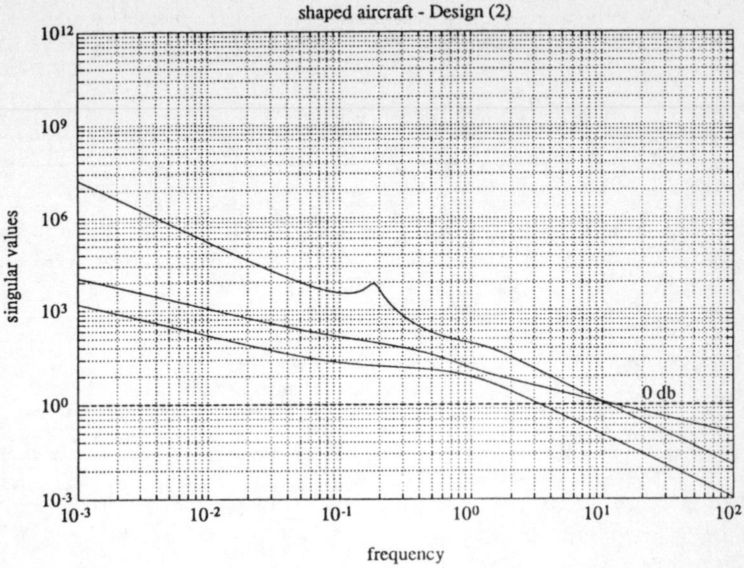

Figure 7.26 Open-Loop Singular Values–Shaped Plant (2)

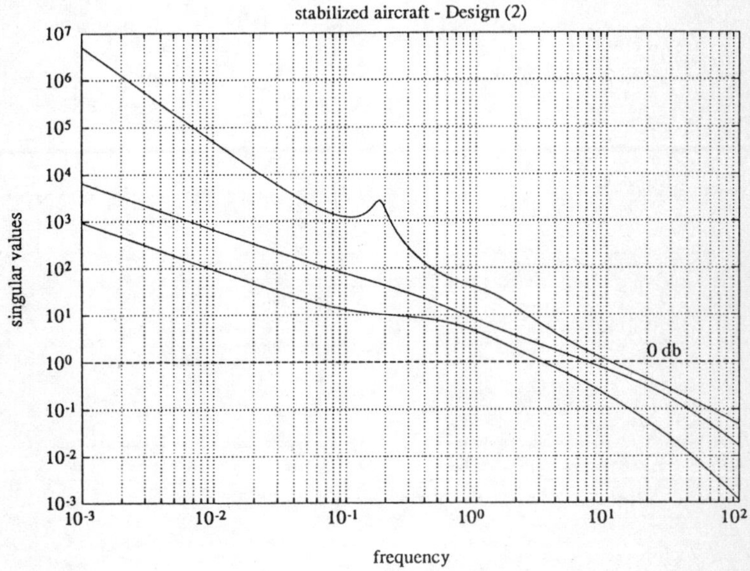

Figure 7.27 Open-Loop Singular Values–Stabilized Plant (2)

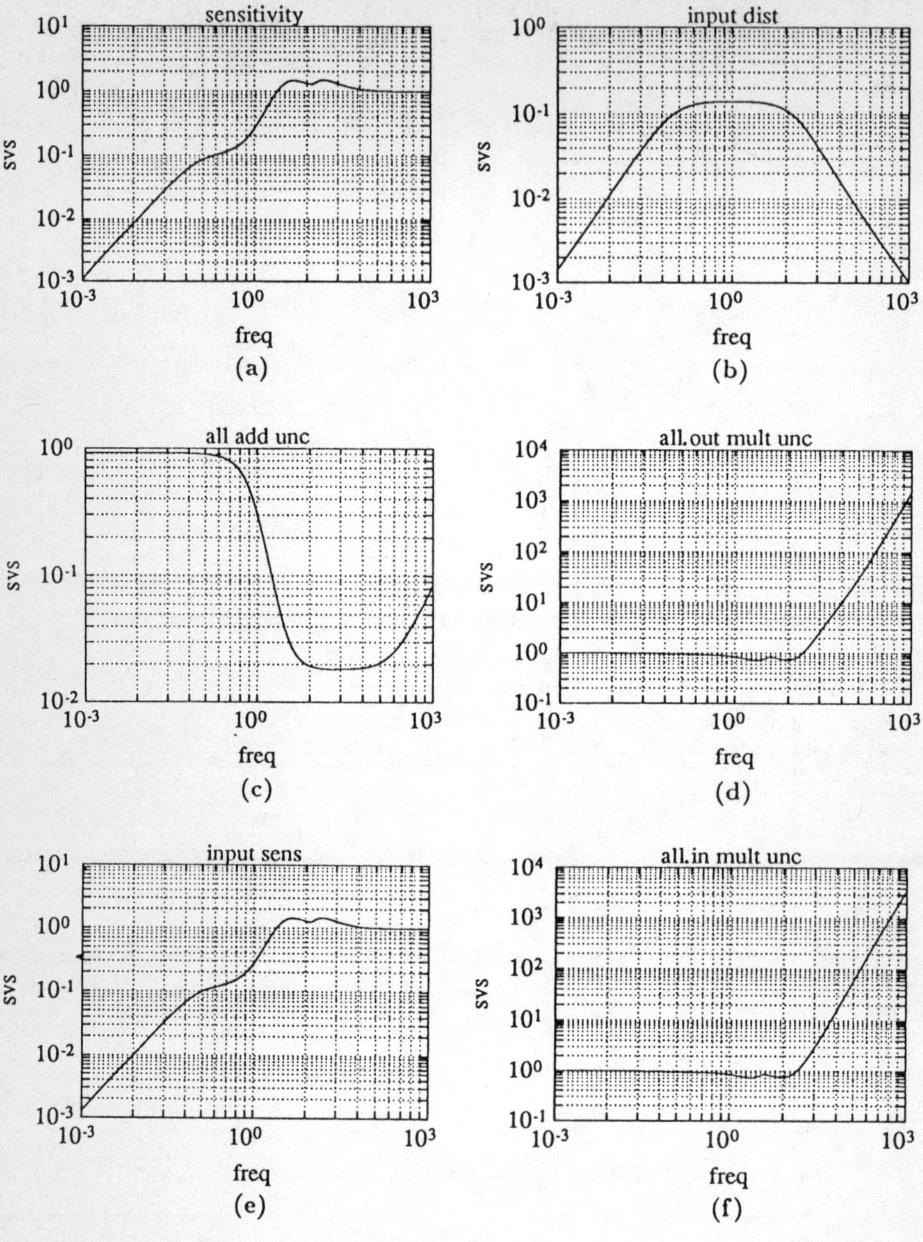

Figure 7.28 Frequency Domain Indicators–Design (2)

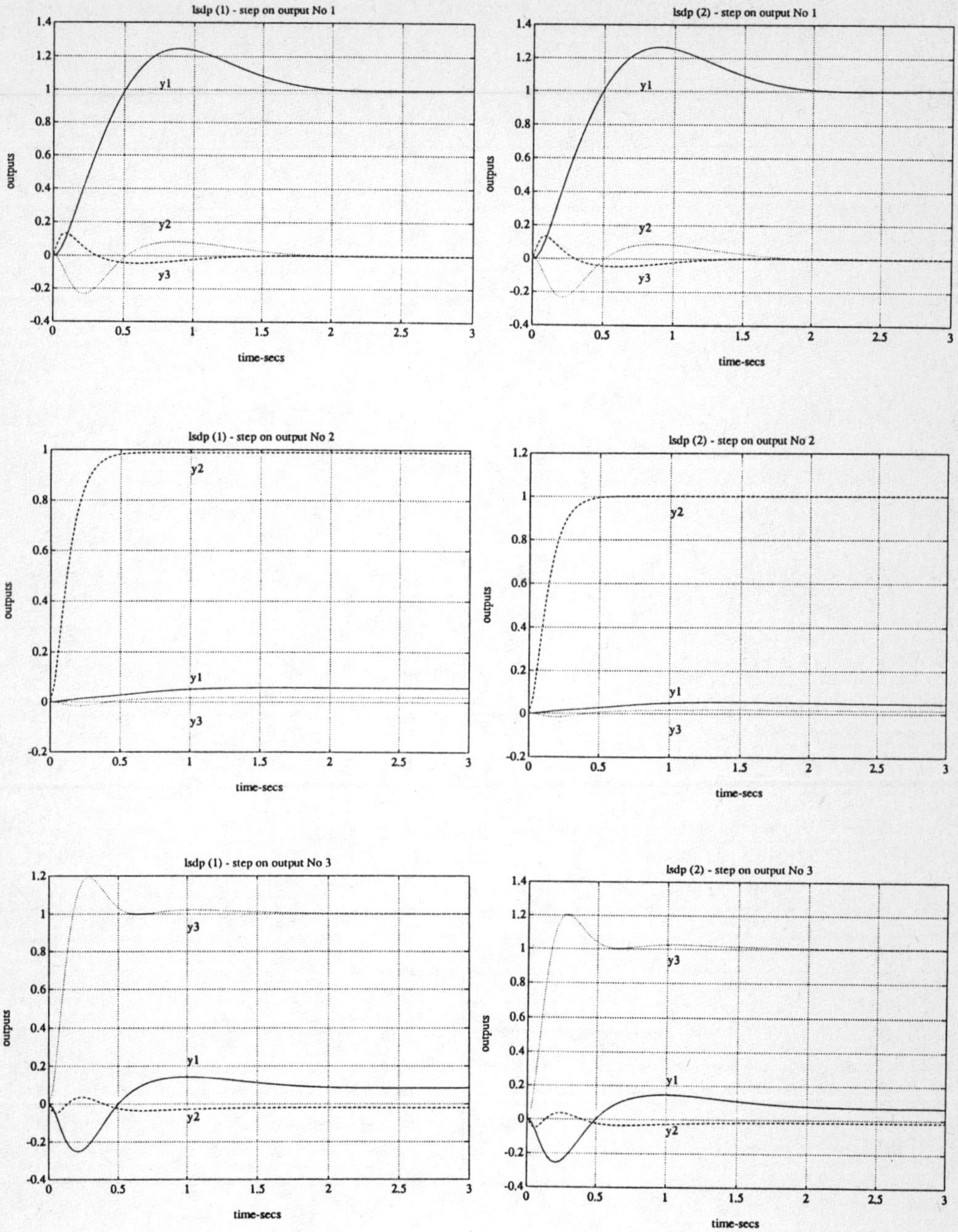

Figure 7.29 Output Responses to Input Steps - Designs (1) & (2)

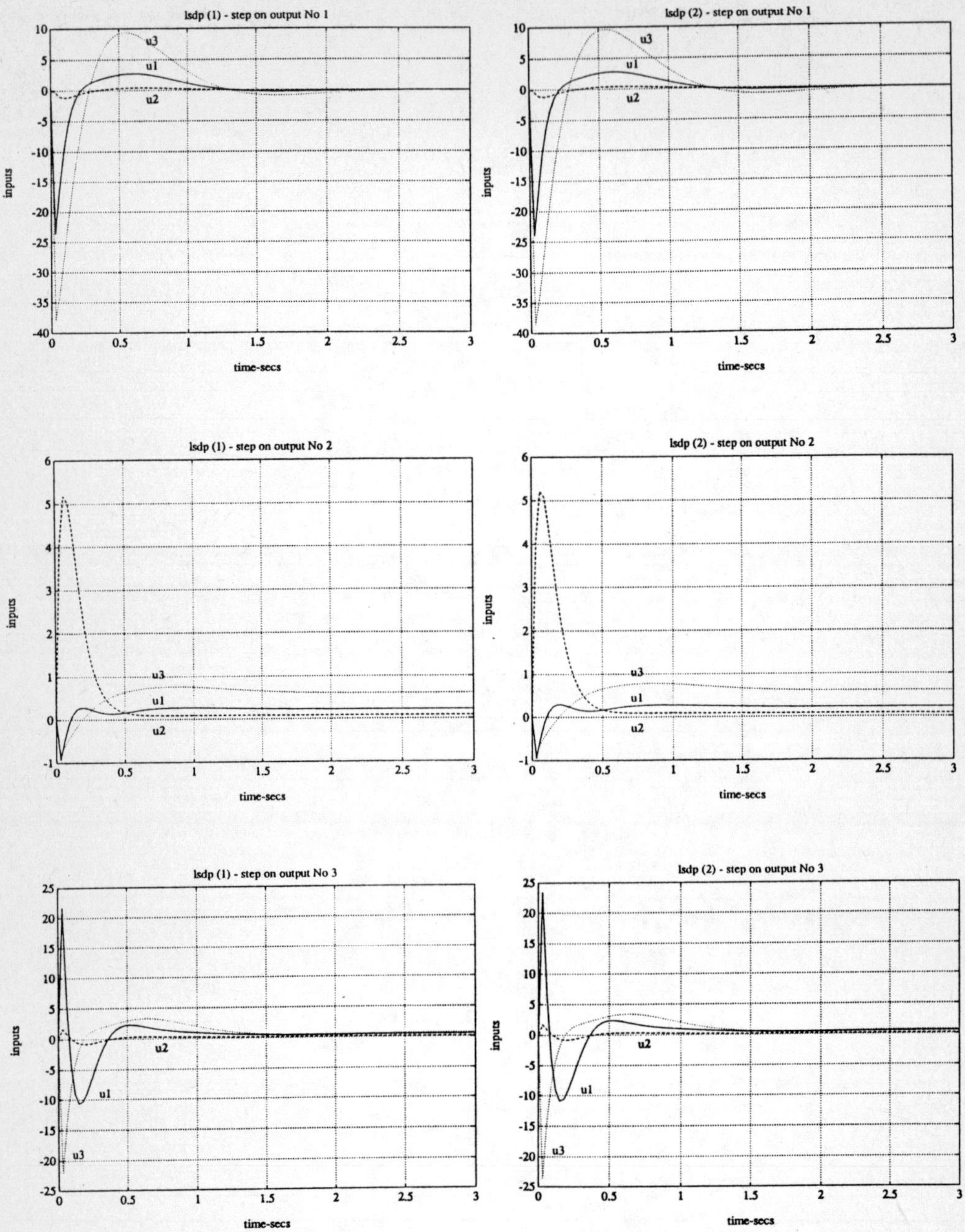

Figure 7.30 Input Responses to Input Steps - Designs (1) & (2)

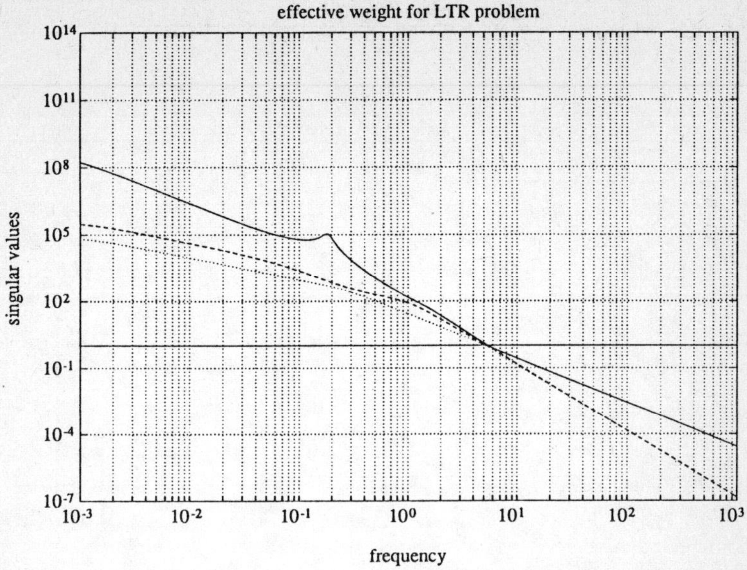

Figure 7.31 Effective Weight for LTR Problem

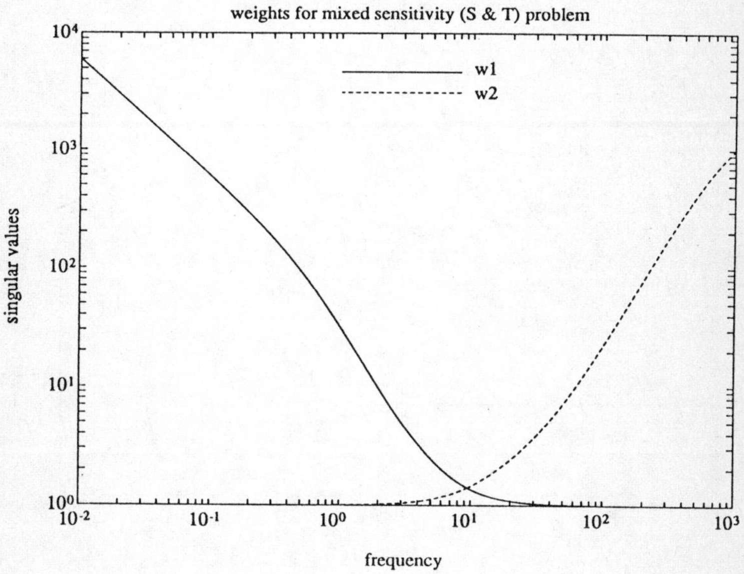

Figure 7.32 Weights for Mixed Sensitivity Problem

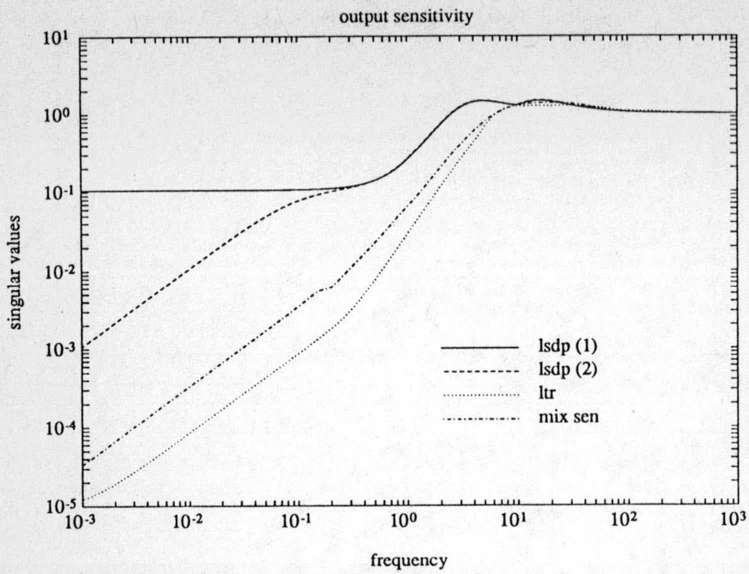

Figure 7.33 $\overline{\sigma}((I - GK)^{-1})$ **V's Frequency**

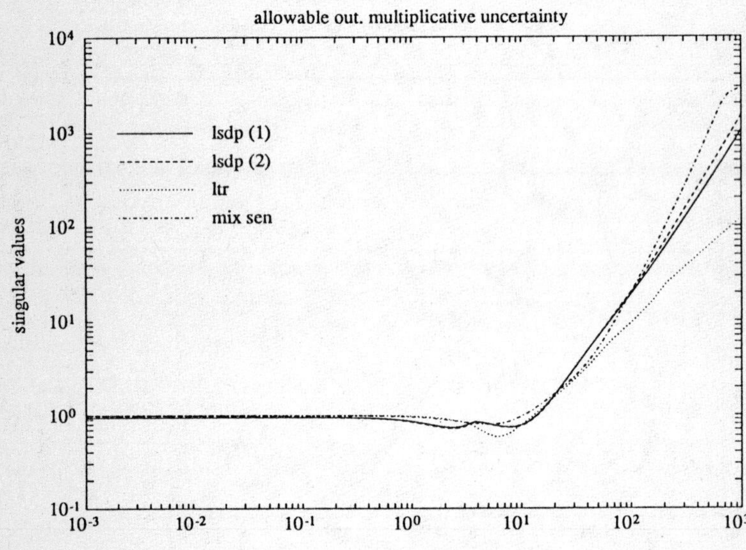

Figure 7.34 $1/(\overline{\sigma}(GK(I - GK)^{-1}))$ **V's Frequency**

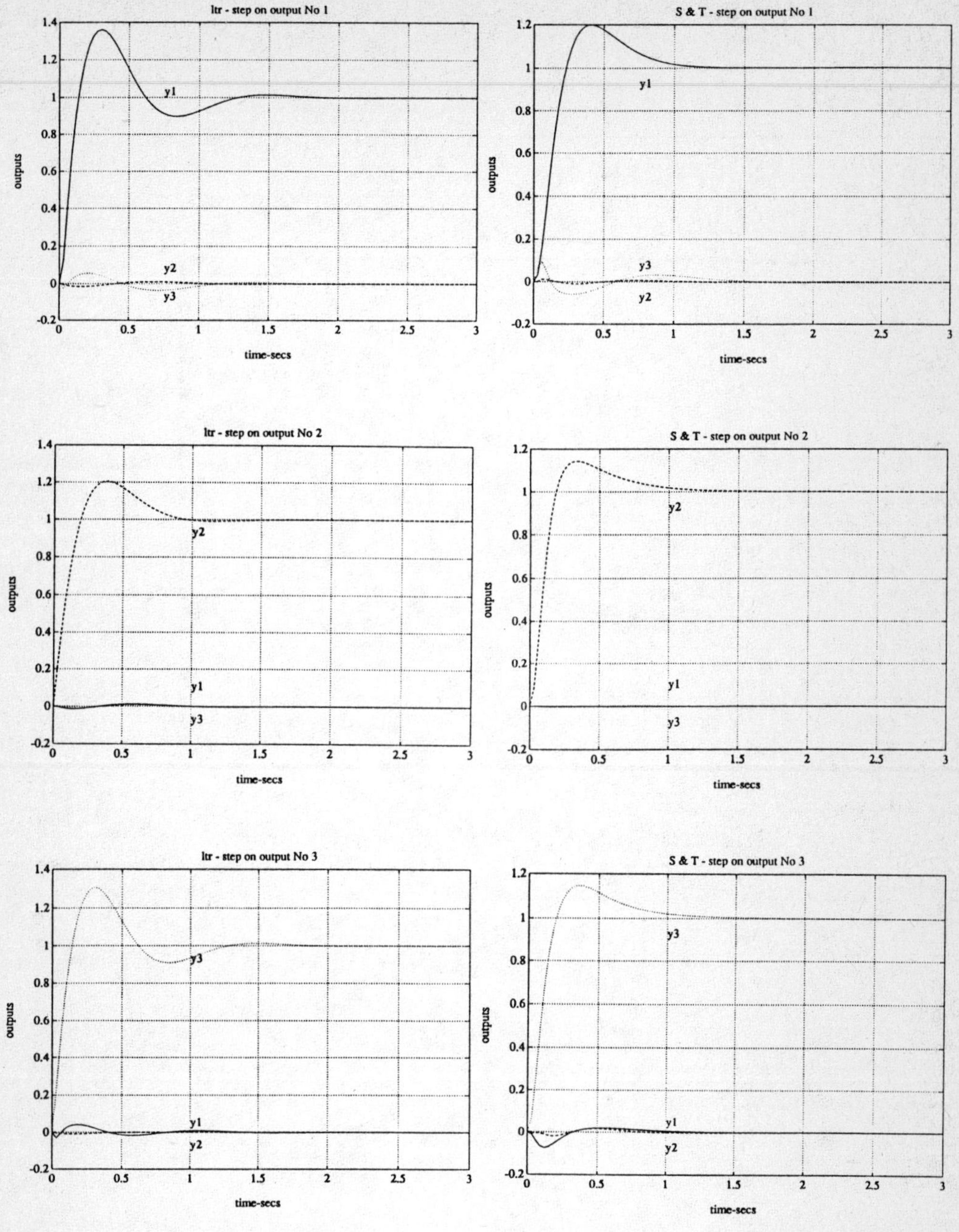

Figure 7.35 Output Responses to Input Steps - LTR and S & T Designs

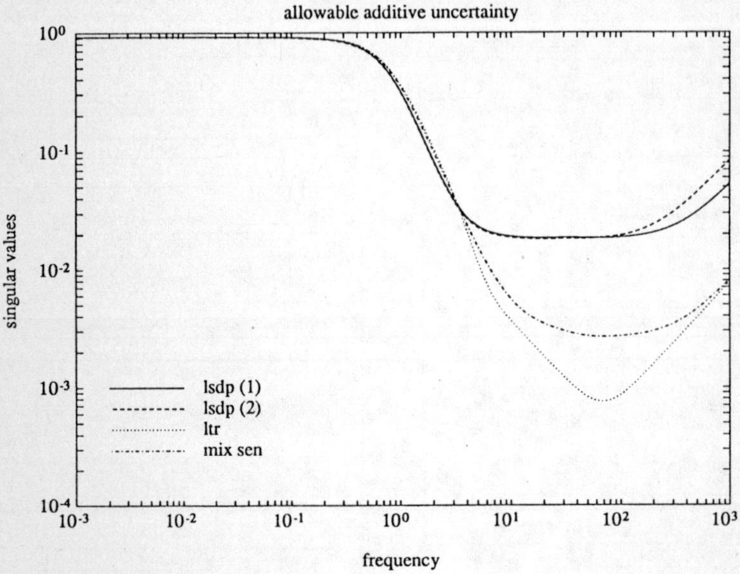

Figure 7.36 $1/(\overline{\sigma}(K(I-GK)^{-1}))$ **V's Frequency**

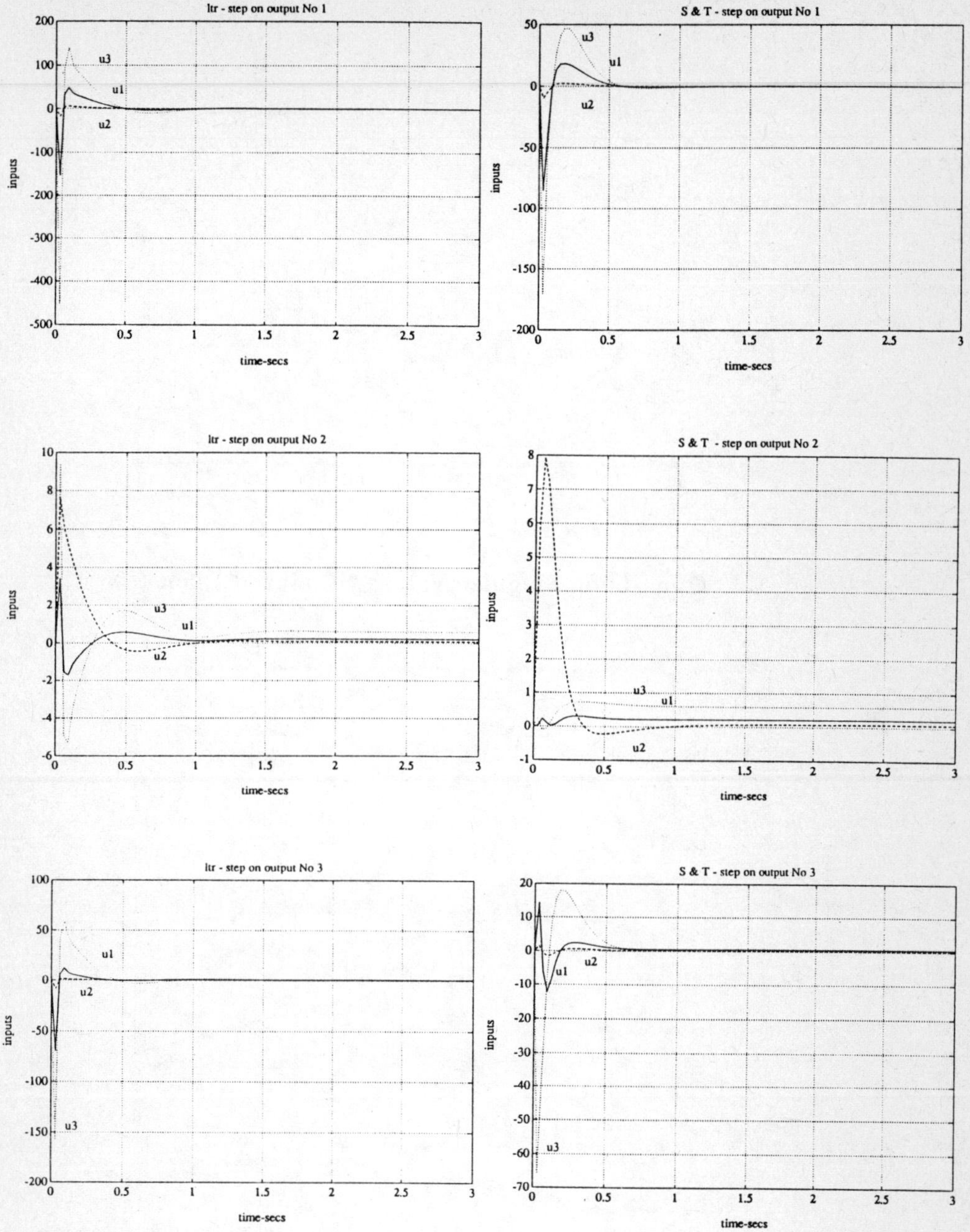

Figure 7.37 Input Responses to Input Steps - LTR and S& T Designs

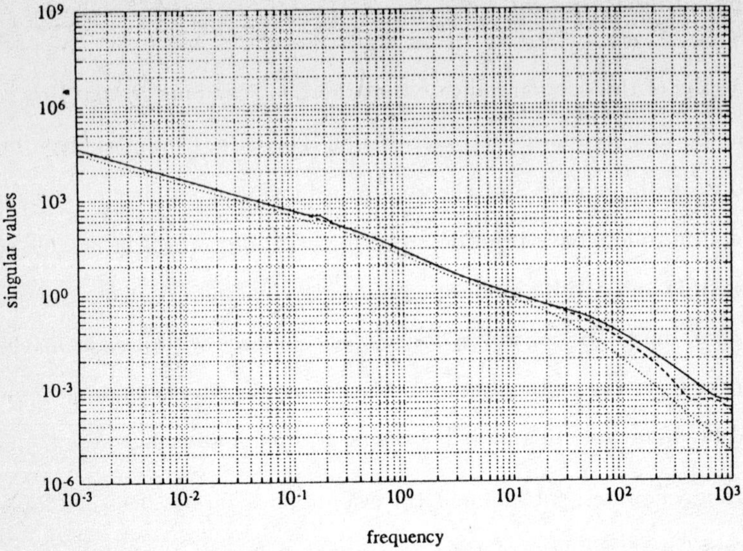

Figure 7.38 Open-Loop Singular Values–Stabilized Plant S & T

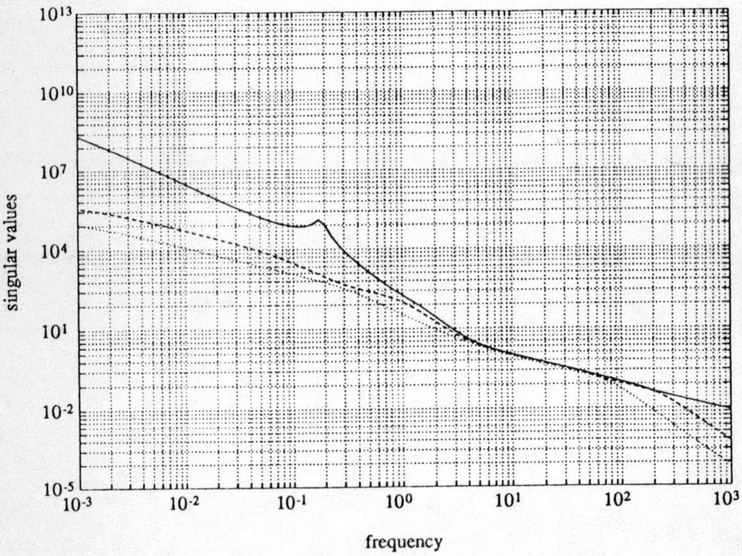

Figure 7.39 Open-Loop Singular Values–Stabilized Plant LTR

7.5 Concluding Remarks

The results of this section indicate that the Loop Shaping Design Procedure is an effective method for feedback controller design. The three examples given have shown the procedure to be simple and effective in achieving a number of specified design objectives. Controllers have been designed for attitude control of flexible spacecraft, using both SISO and MIMO models, and vertical dynamics control of an aircraft. In the latter case, the LSDP controller design was compared with designs using the LQG Loop Transfer Recovery method and the H_∞ Mixed Sensitivity approach on the same model, and fundamental differences between the three approaches were highlighted.

As with any new design technique, the evaluation so far has been of a limited scope. To strengthen the credibility of the Loop Shaping Design Procedure it is therefore necessary to apply the procedure to a number of further designs. A design on a noted 'benchmark' system would be of particular value. In addition, the selection of more complicated shaping functions to enable 'tighter' control over individual loops of a MIMO system needs to be investigated.

APPENDIX A

The Algebraic Riccati Equation

Consider the state-space model of the form

$$\left.\begin{aligned} \dot{x} &= Ax + Bu \\ y &= Cx + Du \end{aligned}\right\} \tag{A.1}$$

where $x \in \Re^n$, $y \in \Re^m$, $u \in \Re^r$ and A, B, C, D, are time-invariant matrices of compatible sizes. This system is denoted (A,B,C,D) and in all further results, it is assumed to be a minimal realization of its transfer matrix:

$$G(s) \triangleq C(sI - A)^{-1}B + D \tag{A.2}$$

The two particular Algebraic Riccati Equations (AREs) of interest in this work are the Generalized Control Algebraic Riccati Equation (GCARE):

$$(A - BS^{-1}D^*C)^*X + X(A - BS^{-1}D^*C) - XBS^{-1}B^*X + C^*R^{-1}C = 0 \tag{A.3}$$

and the Generalized Filtering Algebraic Riccati Equation (GFARE):

$$(A - BD^*R^{-1}C)Z + Z(A - BD^*R^{-1}C)^* - ZC^*R^{-1}CZ + BS^{-1}B^* = 0 \tag{A.4}$$

where

$$R \triangleq (I + DD^*) \tag{A.5}$$

$$S \triangleq (I + D^*D) \tag{A.6}$$

and by inspection $R^{-1} = I - DS^{-1}D^*$,$S^{-1} = I - D^*R^{-1}D$, $DS^{-1} = R^{-1}D$ and $DS = RD$.

Associated with these Riccati equations are the closed-loop control and filtering matrices, defined respectively as:

$$A^c \triangleq A + BF \tag{A.7}$$

$$A^o \triangleq A + HC \tag{A.8}$$

Where F, the control gain, and H, the filter gain are defined:

$$F \triangleq -S^{-1}(D^*C + B^*X) \tag{A.9}$$

$$H \triangleq -(BD^* + ZC^*)R^{-1} \tag{A.10}$$

Noting that the controllability of $(A - BS^{-1}D^*C, BS^{-1/2})$ is uniquely implied by the controllability of (A, B) and that the observability of $(R^{-1/2}C, A - BD^*R^{-1}C)$ is uniquely implied by the observability of (C, A) (both can be shown by simple PBH tests), then the following theorems give sufficient (but not necessary) conditions for the existence and uniqueness of particular solutions to GCARE.

Theorem A.1 (Kalman, 1960)

If (A,B) is completely controllable, and (C,A) is completely observable, then there exists a unique solution, $X = X^* > 0$ to GCARE and the eigenvalues of A^c have strictly negative real parts.

Remark A.2

It should be noted that considerably weaker conditions would be sufficient to yield the solutions stated in Theorem A.1. The condition of minimality is assumed as this is compatible with assumptions made in the rest of the paper.

Remark A.3

Theorem A.1 can be applied directly to GFARE and equivalent results obtained, if the system (A, B, C, D) is replaced by (A^*, C^*, B^*, D^*), X replaced by Z, and hence A^c replaced by A^o.

It can also be shown that X and Z defined in (A.3), (A.4) respectively, solve:

$$(A-BS^{-1}D^*C)^*Z^{-1}+Z^{-1}(A-BS^{-1}D^*C)+Z^{-1}BS^{-1}B^*Z^{-1}-C^*R^{-1}C = 0 \tag{A.11}$$

$$(A - BD^*R^{-1}C)X^{-1} + X^{-1}(A - BD^*R^{-1}C)^* + X^{-1}C^*R^{-1}CX^{-1} - BS^{-1}B^* = 0 \tag{A.12}$$

It is also possible to relate the stabilizing solutions of GCARE and GFARE.

Theorem A.4 (Bucy,[2])

$$A^o = (I + ZX)A^c(I + ZX)^{-1} \tag{A.13}$$

$$(A^o)^* = (I + XZ)^{-1}(A^c)^*(I + XZ). \tag{A.14}$$

■

(These were proven by Bucy, (1972) for the case $D = 0$, but can be readily extended to the $D \neq 0$ case as well.)

Finally for completeness, the stabilizing solutions of GCARE and GFARE can be shown to satisfy the following related Lyapunov equations:

$$\begin{aligned} XA^c + (A^c)^*X &= -(C + DF)^*(C + DF) - F^*F \qquad (A.15) \\ &= -C^*R^{-1}C - XBS^{-1}B^*X \end{aligned}$$

$$\begin{aligned} A^oZ + Z(A^o)^* &= -(B + HD)(B + HD)^* - HH^* \qquad (A.16) \\ &= -BS^{-1}B^* - ZC^*R^{-1}CZ \end{aligned}$$

These are a direct result of (A.3) and (A.4). A further two Lyapunov equations can be obtained by combining (A.3) with (A.14) and (A.4) with (A.14):

$$(Z^{-1} + X)^{-1}(A^c)^* + A^c(Z^{-1} + X)^{-1} = -BS^{-1}B^* \tag{A.17}$$

$$(X^{-1} + Z)^{-1}A^o + (A^o)^*(X^{-1} + Z)^{-1} = -C^*R^{-1}C \tag{A.18}$$

APPENDIX B

Suboptimal Nehari Extensions

A state-space characterization will be derived here for all sub-optimal extensions of an unstable function that is constrained to satisfy an 'inner' requirement. We firstly state a more general result characterizing all sub-optimal extensions of any unstable function. This is derived from Glover, (1987).

Lemma B.1

All sub-optimal extensions of a function $\boldsymbol{R}$, $\boldsymbol{R}^* \in RH_\infty^{m\times p}$, of degree n, with state-space form $R = (A, B, C, D)$, given by

$$\|\boldsymbol{R} + \boldsymbol{Q}\|_\infty \leq \alpha$$

can be written $\boldsymbol{Q} \in RH_\infty^{p\times m}$, where

$$\boldsymbol{Q} = \mathcal{T}_{\boldsymbol{V}}[\phi], \qquad (B.1)$$

$\phi \in RH_\infty^{p\times m}$, $\|\phi\|_\infty \leq 1$, and

$$\boldsymbol{V} = \left[\begin{array}{c:c} V_{11} & V_{12} \\ \hdashline V_{21} & V_{22} \end{array}\right] = \left[\begin{array}{c|c:c} -A^* & W^{*-1}C^* & \alpha^{-1}W^{*-1}QB \\ \hline -(CP + DB^*) & I & -\alpha^{-1}D \\ \hdashline B^* & 0 & \alpha^{-1}I \end{array}\right] \qquad (B.2)$$

where -Q (resp -P) is the controllability (resp) observability Gramian of $\boldsymbol{R}^*$, and $W \triangleq (PQ - \alpha^2 I)$. ■

We now characterize all sub-optimal extensions of an unstable function $\boldsymbol{R}$ satisfying $\boldsymbol{R}^*\boldsymbol{R} = I$. (That is $\boldsymbol{R}^*$ is co-inner.)

Lemma B.2

Given a co-inner function $\boldsymbol{R}^* \in RH_\infty^{m\times p}$, $m \leq p$, of degree n, with $\boldsymbol{R}$ having state-space realization $\boldsymbol{R} = (A, B, C, D)$, then all transfer functions $\boldsymbol{Q} \in RH_\infty^{p\times m}$, achieving

$$\|\boldsymbol{R} + \boldsymbol{Q}\|_\infty \leq \alpha \qquad (B.3)$$

can be written

$$Q = \mathcal{T}_U[\Phi] \tag{B.4}$$

where Φ is an arbitrary transfer function constrained to satisfy $\Phi \in RH_\infty^{p\times m}$, $\|\Phi\|_\infty \le 1$, and

$$U = \left[\begin{array}{cc:c} U_{11} & -\alpha U_{12} & U_{12} \\ \hdashline U_{21} & I - \alpha U_{22} & U_{22} \end{array}\right] \tag{B.5}$$

with state-space form

$$\left[\begin{array}{c:c} U_{11} & U_{12} \\ \hdashline U_{21} & U_{22} \end{array}\right] = \left[\begin{array}{c|c:c} -A^* & W^{*-1}C^*D_\perp & \alpha^{-1}W^{*-1}QB \\ \hline -(CP+DB^*) & D_\perp & -\alpha^{-1}D \\ \hdashline B^* & 0 & \alpha^{-1}I \end{array}\right] \tag{B.6}$$

and -Q (resp -P) is the controllability (resp observability) Gramian of R^*, $W \triangleq (PQ - \alpha^2 I)$, and $D_\perp$ is the unitary completion of D i.e. $[D_\perp, D]$ is square unitary.

Proof

Noting that in Lemma B.1 that ϕ is an *arbitrary* contraction, V_{11}, V_{21} can be postmultiplied by a unitary matrix without changing the parametrization in (B.1). Next note that $R^*R = I \Rightarrow D^*D = I$. Hence, if the unitary completion of D, $D_\perp$, is chosen, so that the matrix

$$S = [D_\perp, D] \tag{B.7}$$

is unitary, i.e. $SS^* = S^*S = I$, then a matrix U, where

$$U = \left[\begin{array}{c:c} U_{11} & U_{12} \\ \hdashline U_{21} & U_{22} \end{array}\right] = \left[\begin{array}{c:c} V_{11}S & V_{12} \\ \hdashline V_{21}S & V_{22} \end{array}\right]$$

will also parametrize all Q in (B.1) and $\Phi = S^{-1}\phi$ again satisfies $\|\Phi\|_\infty \le 1$. Note from (B.2) that

$$\begin{bmatrix} U_{11} \\ U_{21} \end{bmatrix} = \begin{bmatrix} V_{11} \\ V_{21} \end{bmatrix} S = \left[\begin{array}{c|cc} -A^* & W^{*-1}C^*D_\perp & W^{*-1}C^*D \\ \hline -(CP+DB^*) & D_\perp & D \\ B^* & 0 & 0 \end{array}\right]. \tag{B.8}$$

Further, R^* has a state-space realization given by $(-A^*, C^*, -B^*, D^*)$, and noting from Proposition 2.8 that for a co-inner function $C^*D = -QB$, then the result in (B.6) is immediate. ■

APPENDIX C

Proof of Miscellaneous Results

1. Proof of Lemma 4.4

The proof is straightforward by noting the following Lemma due to Redheffer:

Lemma C.1 (Redheffer, 1960)

For $\boldsymbol{J}, \boldsymbol{K} \in RL_\infty$ with $\|\boldsymbol{J}\|_\infty \leq \sigma$, $\|\boldsymbol{J}_{22}\boldsymbol{K}\|_\infty < 1$ then $\|\mathcal{F}_L(\boldsymbol{J}, \boldsymbol{K})\|_\infty \leq \sigma$ if $\|\boldsymbol{K}\|_\infty \leq \sigma^{-1}$. ■

Proof of Lemma 4.4: First, to prove (a), let $\boldsymbol{F}$ be defined as

$$\boldsymbol{F} = \mathcal{F}_L(\boldsymbol{J}, \begin{bmatrix} E_1 \\ E_2 \end{bmatrix})$$

where

$$\boldsymbol{J} = \left[\begin{array}{c|cc} 0 & \alpha^{-1}I & 0 \\ \hline \sqrt{\alpha^{-2}-1}I & 0 & I \end{array}\right]$$

Noting that as $\boldsymbol{J}\boldsymbol{J}^* = \alpha^{-2}I$ then

$$\|\boldsymbol{F}\| \leq \alpha^{-1} \quad \text{(by Lemma C.1)}.$$

Now, by Definition 2.13,

$$\boldsymbol{F} = \alpha^{-1}\sqrt{\alpha^{-2}-1}\;\boldsymbol{E}_1(I - \boldsymbol{E}_2)^{-1}$$

$$\Rightarrow \|\boldsymbol{E}_1(I - \boldsymbol{E}_2)^{-1}\|_\infty \leq \alpha(1-\alpha^2)^{-1/2}.$$

To prove (b), note that the selection $\boldsymbol{E}_1 = (1-\alpha^2)\boldsymbol{F}$, $\boldsymbol{E}_2 = \alpha I$ demonstrates that for any $\boldsymbol{F} \in RH_\infty$ satisfying $\|\boldsymbol{F}\|_\infty \leq \alpha(1-\alpha^2)^{-1/2}$ there exists $\boldsymbol{E}_1$, $\boldsymbol{E}_2 \in RH_\infty$ such that $\boldsymbol{E}_1(I - \boldsymbol{E}_2)^{-1} = \boldsymbol{F}$ satisfying

$$\left\|\begin{bmatrix} \boldsymbol{E}_1 \\ \boldsymbol{E}_2 \end{bmatrix}\right\|_\infty \leq \alpha.$$

■

2. Proof of Lemma 4.5

The proof of this Lemma uses a well known result from Hankel operator theory:

Lemma C.2 (Francis,1987,p70)

Let

$$\|\boldsymbol{R}-\boldsymbol{X}\|_\infty = \|\boldsymbol{R}^*\|_H \triangleq \sigma_1(\boldsymbol{R}^*) \tag{C.1}$$

where $\boldsymbol{R}^*$, $\boldsymbol{X} \in RH_\infty$. Then, there exists vectors $g(s)$ and $f(s) \in RH_2$, independent of $\boldsymbol{X}$, such that

$$(\boldsymbol{R}-\boldsymbol{X})g(s) = \sigma_1(\boldsymbol{R}^*)f(-s) \tag{C.2}$$

■

Proof of Lemma 4.5:

(i) It is well known that $\|[\tilde{\boldsymbol{N}},\tilde{\boldsymbol{M}}]\|_H \leq \|[\tilde{\boldsymbol{N}},\tilde{\boldsymbol{M}}]\|_\infty = 1$.

(ii) Suppose $\|[\tilde{\boldsymbol{N}},\tilde{\boldsymbol{M}}]\|_H = 1$ then from Lemma C.2, there exists $g(s)$,$f(s) \in RH_2$ such that

$$\begin{bmatrix} \tilde{\boldsymbol{N}}^* \\ \tilde{\boldsymbol{M}}^* \end{bmatrix} g(s) = f(-s) \tag{C.3}$$

(i.e. $\boldsymbol{R} = \begin{bmatrix} \tilde{\boldsymbol{N}}^* \\ \tilde{\boldsymbol{M}}^* \end{bmatrix}$, and $\boldsymbol{X} = 0$ in Lemma C.2)

But, one of the requirements for a coprime factorization (see Def. 2.15) is that there exists $\boldsymbol{U}$, $\boldsymbol{V} \in RH_\infty$:

$$\boldsymbol{U}^*\tilde{\boldsymbol{N}}^* + \boldsymbol{V}^*\tilde{\boldsymbol{M}}^* = I \tag{C.4}$$

premultiplying (C.3) by $[\boldsymbol{U}^*,\boldsymbol{V}^*]$ yields

$$g(s) = [\boldsymbol{U}^*,\boldsymbol{V}^*]f(-s)$$

which is a contradiction as the right hand side $\notin RH_2$ ■

3. Proof of Theorem 4.26

The proof here follows that of Opdenacker and Jonckheere, (1988, Theorem 2), where we extend the result of that theorem to the case where $D \neq 0$.

To begin, let $G \stackrel{s}{=} (A, B, C, D)$ represent the passive system in question. Next we define the so called 'scattering system' associated with G by $S \stackrel{s}{=} (\tilde{A}, \tilde{B}, \tilde{C}, \tilde{D})$ where

$$\tilde{A} = A - B(I + D)^{-1}C \tag{C.5}$$

$$\tilde{B} = \sqrt{2}B(I + D)^{-1} \tag{C.6}$$

$$\tilde{C} = \sqrt{2}(I + D)^{-1}C \tag{C.7}$$

$$\tilde{D} = (D - I)(I + D)^{-1}. \tag{C.8}$$

Then by Opdenacker and Jonckheere, (1988), we have the following result.

Lemma C.3 (Opdenacker and Jonckheere, 1988)

The following statements are equivalent

(i) G is positive real (hence stable).

(ii) S is bounded real. i.e. S is asymptotically stable and $\|S\|_\infty \leq 1$.

We next state an elementary result for bounded real systems.

Lemma C.4

Let $(\tilde{A}, \tilde{B}, \tilde{C}, \tilde{D})$ be a state-space realization of a bounded real system S, and let P (resp. Q) be the associated controllability (resp. observability) Gramian. Then P and Q solve

$$P\tilde{A}^* + \tilde{A}P + \tilde{B}\tilde{B}^* = 0 \tag{C.9}$$

$$Q\tilde{A} + \tilde{A}^*Q + \tilde{C}^*\tilde{C} = 0 \tag{C.10}$$

and

$$\lambda_{max}(PQ) \leq 1. \tag{C.11}$$

Proof:

This is immediate by noting that

$$\lambda_{max}(PQ) = \|S\|_H \leq \|S\|_\infty = 1.$$

■

We now return to the proof of Theorem 4.31. Firstly, note that as $\boldsymbol{S}$ is bounded real (and hence asymptotically stable) we have from (C.5) that

$$Re[\lambda_i(\tilde{A})] = Re[\lambda_i(A - B(I + D)^{-1}C)] < 0$$

for all $i = 1, \ldots, n$. This implies that a state feedback of the form

$$F_s = -(I + D)^{-1}C \tag{C.12}$$

applied to $\boldsymbol{G}$ would be stabilizing.

Next, following the proof in Opdenacker and Jonckheere, (1988), we consider the Linear Quadratic Regulator (LQR) problem of minimizing the cost criterion

$$\mathcal{J} = \int_0^\infty (y(t)^* y(t) + u(t)^* u(t))dt$$

subject to the conditions $\dot{x}(t) = Ax(t) + Bu(t)$, $x(0) = x_o$, $y(t) = Cx(t) + Du(t)$. The optimal cost for this problem, denoted $\mathcal{J}_o$, is given by

$$\mathcal{J}_o = x_o^* X x_o, \qquad X > 0, \tag{C.13}$$

where X solves GCARE in (A.3). By (A.15), this equation can be rewritten as

$$X(A + BF) + (A + BF)^* X + (C + DF)^*(C + DF) + F^* F = 0 \tag{C.14}$$

and

$$F = -S^{-1}(D^* C + B^* X). \tag{C.15}$$

We also note that, as F_s in (C.12) is a stabilizing feedback, by Kwakernnak and Sivan, (1972, p322), this can be shown to have an associated LQR cost,

$$\mathcal{J}_s = x_o^* X_s x_o \tag{C.16}$$

where X_s is the positive definite solution to

$$X_s(A+BF_s)+(A+BF_s)^*X_s+(C+DF_s)^*(C+DF_s)+F_s^*F_s=0. \qquad (C.17)$$

By (C.12) we can rewrite (C.17) as

$$\begin{aligned}&X_s(A-B(I+D)^{-1}C)+(A-B(I+D)^{-1}C)^*X_s\\&+(C-D(I+D)^{-1}C)^*(C-D(I+D)^{-1}C)+C^*(I+D)^{*-1}(I+D)^{-1}C=0\\&\Rightarrow X_s\tilde{A}+\tilde{A}^*X_s+\tilde{C}^*\tilde{C}=0.\end{aligned}$$

Hence comparing with (C.10), we can see that $X_s=Q$, the observability Gramian of the bounded real system S. Further, noting that as $\mathcal{J}_o\leq\mathcal{J}_s$, we have that

$$X\leq X_s. \qquad (C.18)$$

Similarly, if we were to apply an identical analysis to the dual LQR cost criterion

$$\mathcal{J}_1=\int_0^\infty(y_1(t)^*y_1(t)+u_1(t)^*u_1(t))dt$$

subject to the conditions $\dot{x}_1(t)=A^*x_1(t)+C^*u_1(t)$, $x_1(0)=x_{1o}$, $y_1(t)=B^*x_1(t)+D^*u_1(t)$, it can be shown that the optimal cost is achieved using Z, the solution to GFARE, and that the inequality

$$Z\leq Z_s \qquad (C.19)$$

holds, where $Z_s=P$ is the controllability Gramian of the bounded real system S.

Then, by Lemma C.4, (C.18), and (C.19) we have that

$$\mu_i\triangleq\lambda_i(XZ)\leq\lambda_i(X_sZ_s)\leq 1$$

for all $i=1,\ldots,n$. This completes the proof of Theorem 4.31. ■

APPENDIX D

State-Space Systems for Chapter 7

D.1 State-Space Matrices for Design Example 1

1. Controller for Design (1) - K_1

```
A=

-23.7320     1.0000  -23.7320         0
 -4.4097    -0.2308   -4.3847   -0.0212
 -8.4894          0   -8.4894    1.0000
-19.3781    -5.0461  -21.1999   -0.4724

B=

1.0e+04 *

4.7464
0.8750
1.6979
3.7242

C=

1.0000    6.6643    0.2780    0.6117

D=

0
```

2. Controller for Design (2) - K_2

```
A=

0    10.6591    -4.3091    -3.3354     -7.8802
0  -100.5513    72.1135   -73.5563   -299.0902
0     0.6352    -0.3807     1.0774      2.0607
0     0.3630    -0.6036    -0.1976     -0.7720
0     0.7674    -0.6002    -0.4013     -1.6053

B=

       0
384.8650
 -1.3706
  0.5944
  0.7300

C=

1.0e+03 *

0.0200    5.3296    -2.1545    -1.6677    -3.9401

D=

0
```

D.2 State-Space Matrices for Design Example 2

1. Nominal Space Platform Plant

```
A=

Columns 1 through 7
     0    1.0000         0         0         0         0         0
     0         0         0         0         0         0         0
     0         0         0    1.0000         0         0         0
     0         0         0         0         0         0         0
     0         0         0         0         0    1.0000         0
     0         0         0         0   -0.1187   -0.0069         0
     0         0         0         0         0         0         0
     0         0         0         0         0         0   -0.2819
     0         0         0         0         0         0         0
     0         0         0         0         0         0         0

Columns 8 through 10
     0         0         0
     0         0         0
     0         0         0
     0         0         0
     0         0         0
     0         0         0
1.0000         0         0
-0.0106        0         0
     0         0    1.0000
     0   -0.5805   -0.0152

B=
     0         0         0         0
0.0017   -0.0010   -0.0131    0.0002
     0         0         0         0
0.0018    0.0011    0.0064    0.0067
     0         0         0         0
0.0013    0.0005   -0.0114    0.0088
     0         0         0         0
0.0028    0.0003    0.0058    0.0038
     0         0         0         0
0.0038   -0.0015    0.0025   -0.0063

C=

Columns 1 through 7
 0.0017         0    0.0018         0    0.0013         0    0.0028
-0.0010         0    0.0011         0    0.0005         0    0.0003
      0    0.0017         0    0.0018         0    0.0013         0
      0   -0.0010         0    0.0011         0    0.0005         0
      0    0.0017         0    0.0018         0    0.0064         0

Columns 8 through 10
      0    0.0038         0
      0   -0.0015         0
 0.0028         0    0.0038
 0.0003         0   -0.0015
-0.0006         0   -0.0055

D=

  0   0   0   0
  0   0   0   0
  0   0   0   0
  0   0   0   0
  0   0   0   0
```

2. Perturbed Space Platform Plant

```
A=

Columns 1 through 7
       0    1.0000         0         0         0         0         0
       0         0         0         0         0         0         0
       0         0         0    1.0000         0         0         0
       0         0         0         0         0         0         0
       0         0         0         0         0    1.0000         0
       0         0         0         0   -0.1300   -0.0036         0
       0         0         0         0         0         0         0
       0         0         0         0         0         0   -0.2876
       0         0         0         0         0         0         0
       0         0         0         0         0         0         0

Columns 8 through 10
       0         0         0
       0         0         0
       0         0         0
       0         0         0
       0         0         0
       0         0         0
  1.0000         0         0
 -0.0054         0         0
       0         0    1.0000
       0   -0.6387   -0.0080

B=

       0         0         0         0
  0.0008   -0.0012   -0.0135   -0.0021
       0         0         0         0
  0.0021    0.0006    0.0011    0.0059
       0         0         0         0
  0.0016    0.0004   -0.0109    0.0086
       0         0         0         0
  0.0025    0.0004    0.0062    0.0037
       0         0         0         0
  0.0038   -0.0016    0.0033   -0.0064

C=

Columns 1 through 7
  0.0008         0    0.0021         0    0.0016         0    0.0025
 -0.0012         0    0.0006         0    0.0004         0    0.0004
       0    0.0008         0    0.0021         0    0.0016         0
       0   -0.0012         0    0.0006         0    0.0004         0
       0    0.0008         0    0.0021         0    0.0060         0

Columns 8 through 10
       0    0.0038         0
       0   -0.0016         0
  0.0025         0    0.0038
  0.0004         0   -0.0016
 -0.0005         0   -0.0058

D=

     0     0     0     0
     0     0     0     0
     0     0     0     0
     0     0     0     0
     0     0     0     0
```

3. Controller (4 Plant Inputs) - K_3

A=

Columns 1 through 7

-9.4152	-0.0402	-24.4362	-11.8265	-5.5490	17.7206	-4.3232
0.0123	-1.2286	-0.2333	-0.8251	0.7558	0.7295	-1.2174
0.0920	0.1171	-0.6752	-0.5095	-0.1901	-0.3531	-0.4163
-0.0479	0.0953	0.4523	-0.1250	-0.3075	-0.0484	-0.1107
0.0489	-0.0101	0.0403	0.2887	-0.0520	0.1410	0.0666
0.1149	0.0338	-0.3611	0.1867	-0.2583	-0.3844	-0.0593
0.0534	0.0096	-0.2121	0.1220	-0.0755	-0.3125	-0.1647
-0.0143	-0.0931	0.3833	0.0731	-0.0177	0.0854	0.5616
0.0126	-0.1144	0.0355	0.0983	-0.0121	-0.0149	-0.1109
0.0599	0.0105	-0.0902	0.0560	-0.0531	-0.1986	-0.1346
0	0	0	0	0	0	0
0	0	0	0	0	0	0

Columns 8 through 12

-36.3849	-11.5446	-38.4824	-123.5177	150.2042
-1.1755	3.2577	0.3765	-8.6696	-8.1468
0.0363	0.0633	-0.3457	-0.7840	0.4844
-0.3636	0.1915	-0.2169	-0.4882	0.0376
-0.0206	-0.0550	-0.1206	-0.0121	0.2407
0.4905	0.0650	0.0627	0.0933	-0.0898
-0.3039	0.2274	0.0370	-0.1706	-0.0911
-0.3865	0.3303	-0.3291	-0.2695	0.1559
-0.3149	-0.2715	0.0743	0.1605	0.1634
0.1532	-0.0528	-0.1880	-0.0843	0.0907
0	0	0	0	0
0	0	0	0	0

B=

1.0e+05 *

-5.1877	6.3086	-2.5018	1.9381	-0.7728
-0.3641	-0.3422	-0.1462	-0.0769	-0.0784
-0.0329	0.0204	-0.0094	0.0069	0.0573
-0.0205	0.0016	0.0113	0.0036	-0.0068
-0.0005	0.0101	0.0074	-0.0038	0.0048
0.0039	-0.0038	0.0069	-0.0019	0.0303
-0.0072	-0.0038	-0.0037	0.0004	0.0125
-0.0113	0.0065	0.0031	-0.0045	-0.0155
0.0067	0.0069	-0.0041	-0.0017	-0.0002
-0.0035	0.0038	0.0023	-0.0013	0.0008
0.0042	0	0	0	0
0	0.0042	0	0	0

C=

Columns 1 through 7

-0.0739	-0.1040	-0.0700	0.0521	0.0543	0.0518	-0.0236
0.0219	-0.0113	0.0390	0.0027	-0.0165	-0.0061	0.0102
0.1423	-0.0613	-0.1911	0.1201	-0.0701	-0.2236	-0.1392
-0.0041	-0.1781	0.2375	0.0508	-0.0180	0.1030	0.0758

Columns 8 through 12

0.0573	-0.0867	0.0503	0	0
-0.0406	-0.0060	-0.0122	0	0
0.0643	-0.0766	-0.0910	0	0
-0.2206	-0.1179	0.0187	0	0

D=

0	0	0	0	0
0	0	0	0	0
0	0	0	0	0
0	0	0	0	0

4. Controller (2 Plant Inputs) - Design (1) - K_4

```
A=
Columns 1 through 7
  -0.1979   -0.0092   -0.6512   -0.2228   -0.7871   -0.2439   -0.5546
  -0.0471   -1.2110    4.4231    0.0570   -5.3421   -0.7464   -2.5997
  -0.0218    0.3036   -1.4531    0.9250    1.1961    0.1351    0.5212
  -0.0056    0.0305   -0.1853   -0.2451    0.0901   -0.0286    0.0343
  -0.0478   -0.2973    0.9353   -0.0512   -1.6978    0.7367   -0.8882
  -0.0087   -0.0324    0.0733   -0.0479   -0.2282   -0.2843   -0.1153
  -0.0030   -0.0088    0.0165   -0.0040   -0.0627   -0.0110   -0.0352
  -0.0025   -0.0021   -0.0131   -0.0629   -0.0388   -0.1666   -0.1363
  -0.0060   -0.0037   -0.0267   -0.0108   -0.0627   -0.0149   -0.0408
  -0.0053   -0.0017   -0.0441   -0.2014   -0.0664   -0.2672   -0.0318
  -0.0050    0.0046   -0.0554   -0.0104   -0.0175   -0.0087   -0.0177
  -0.0052    0.0054   -0.0903   -0.4532   -0.0285   -0.1493   -0.0154
        0         0         0         0         0         0         0
        0         0         0         0         0         0         0

Columns 8 through 14
  -0.4061   -1.1635   -0.2208   -1.5145   -0.0853   19.7945    0.9215
  -1.4243   -1.9116    0.0094    6.4576    1.6477    4.7068  121.1035
   0.2904    0.1149   -0.0728   -2.3217   -0.4959    2.1783  -30.3639
  -0.0155   -0.0213   -0.0475   -0.2745   -0.1152    0.3939   -2.9432
  -0.4780   -0.9259   -0.0674    1.1620    0.3844    4.7819   29.7327
  -0.1207   -0.1321   -0.0559    0.0752   -0.0006    0.6884    3.1305
   0.9734   -0.0464    0.0003    0.0088    0.0244    0.3023    0.8841
  -0.0804   -0.0251   -0.0329   -0.0220   -0.0114    0.1193    0.1572
  -0.0335   -0.0756    0.9945   -0.0729    0.0168    0.6039    0.3717
  -0.0855   -0.3327   -0.0941   -0.0607   -0.1069    0.2530    0.1466
  -0.0146   -0.0536   -0.0097   -0.1111    0.9919    0.5035   -0.4647
  -0.0240   -0.0369   -0.1121   -0.6719   -0.2533    0.1380   -0.3850
        0         0         0         0         0         0         0
        0         0         0         0         0         0         0

B=
  1.0e+05 *

   0.8314    0.0387    0.0880    0.0015    0.0457
   0.1977    5.0863   -0.0025    0.3726    0.1947
   0.0915   -1.2753    0.0276   -0.0996   -0.0434
   0.0165   -0.1236    0.0050   -0.0100   -0.0030
   0.2008    1.2488    0.0267    0.0959    0.0620
   0.0289    0.1315    0.0053    0.0104    0.0080
   0.0127    0.0371    0.0004    0.0031    0.0039
   0.0050    0.0066    0.0020    0.0009    0.0050
   0.0254    0.0156    0.0027    0.0015    0.0046
   0.0106    0.0062    0.0074    0.0009    0.0001
   0.0212   -0.0195    0.0039   -0.0012    0.0016
   0.0058   -0.0162    0.0095   -0.0044   -0.0108
   0.0042         0         0         0         0
        0    0.0042         0         0         0

C=
Columns 1 through 7
   0.0100    0.0000    0.1265    0.7800    0.1340    0.8012    0.0509
   0.0000    0.0100   -0.1011   -0.9858    0.1039    0.9723    0.0228

Columns 8 through 14
   0.2574    0.0683    0.2206    0.1075    0.3007         0         0
   0.1026    0.0048    0.0244   -0.0197   -0.1358         0         0

D=
     0     0     0     0     0
     0     0     0     0     0
```

5. Controller (2 Plant Inputs) - Design (2) - K_5

```
A=
Columns 1 through 7
 -0.2152     0.0397    -0.9822    -0.1850    -0.6992    -0.1799    -0.5458
 -0.0625    -4.8685    19.9674     0.4000   -21.8654    -1.7798   -10.3954
 -0.0177     0.9271    -4.0923     0.8873     4.0239     0.3052     1.8581
 -0.0032     0.0681    -0.3246    -0.1926     0.2787    -0.0094     0.1272
 -0.0422    -0.8844     3.4874     0.0351    -4.2883     0.6319    -2.0952
 -0.0053    -0.0691     0.2558    -0.0277    -0.3615    -0.2255    -0.1727
 -0.0016    -0.0122     0.0413    -0.0005    -0.0666    -0.0063    -0.0342
 -0.0011    -0.0030     0.0022    -0.0447    -0.0291    -0.1256    -0.1299
 -0.0027    -0.0047     0.0009    -0.0028    -0.0417    -0.0054    -0.0247
 -0.0024    -0.0023    -0.0170    -0.1528    -0.0412    -0.2039    -0.0159
 -0.0022     0.0069    -0.0445    -0.0033     0.0143    -0.0006     0.0027
 -0.0025     0.0072    -0.0694    -0.3551     0.0040    -0.1198     0.0040
       0          0          0          0          0          0          0
       0          0          0          0          0          0          0

Columns 8 through 14
 -0.3209    -1.3067    -0.1644    -2.1094    -0.0678    21.5200    -3.9688
 -3.0814    -6.5633     0.0150    29.9580     3.9161     6.2502   486.8470
  0.5330     0.9251    -0.0394    -6.3037    -0.7741     1.7674   -92.7067
  0.0194     0.0514    -0.0214    -0.4832    -0.0876     0.2325    -6.7630
 -0.6369    -1.5773    -0.0335     5.0642     0.7077     4.2199    88.4388
 -0.0809    -0.1393    -0.0251     0.3697     0.0351     0.4344     6.8548
  0.9873    -0.0332    -0.0002     0.0543     0.0131     0.1553     1.2203
 -0.0404    -0.0116    -0.0149     0.0068    -0.0047     0.0430     0.2756
 -0.0124    -0.0377     0.9989    -0.0132     0.0088     0.2716     0.4683
 -0.0385    -0.3030    -0.0484    -0.0119    -0.0486     0.1003     0.2135
 -0.0012    -0.0170    -0.0025    -0.0791     0.9943     0.2156    -0.6850
 -0.0085    -0.0075    -0.0517    -0.6457    -0.1283     0.0545    -0.6439
       0          0          0          0          0          0          0
       0          0          0          0          0          0          0

B=
 1.0e+06 *

  0.0452    -0.0083     0.0034    -0.0005     0.0019
  0.0131     1.0224    -0.0014     0.0510     0.0205
  0.0037    -0.1947     0.0010    -0.0098    -0.0036
  0.0005    -0.0142     0.0001    -0.0007    -0.0002
  0.0089     0.1857     0.0005     0.0094     0.0042
  0.0009     0.0144     0.0001     0.0007     0.0004
  0.0003     0.0026     0.0000     0.0001     0.0001
  0.0001     0.0006     0.0000     0.0000     0.0001
  0.0006     0.0010     0.0000     0.0001     0.0001
  0.0002     0.0004     0.0001     0.0000     0.0000
  0.0005    -0.0014     0.0001    -0.0001     0.0000
  0.0001    -0.0014     0.0002    -0.0001    -0.0003
  0.0002          0          0          0          0
       0     0.0002          0          0          0

C=
Columns 1 through 7
  0.0100     0.0000     0.1489     1.2249     0.1583     1.2596     0.0407
  0.0000     0.0100    -0.1235    -1.5542     0.1239     1.4991     0.0148

Columns 8 through 14
  0.2379     0.0478     0.2110     0.0701     0.2915          0          0
  0.0948     0.0019     0.0230    -0.0155    -0.1225          0          0

D=
     0     0     0     0     0
     0     0     0     0     0
```

D.3 State-Space Matrices for Design Example 3

1. Nominal Aircraft Plant

```
A=

        0         0    1.1320         0   -1.0000
        0   -0.0538   -0.1712         0    0.0705
        0         0         0    1.0000         0
        0    0.0485         0   -0.8556   -1.0130
        0   -0.2909         0    1.0532   -0.6859

B=

        0         0         0
  -0.1200    1.0000         0
        0         0         0
   4.4190         0   -1.6650
   1.5750         0   -0.0732

C=

     1     0     0     0     0
     0     1     0     0     0
     0     0     1     0     0

D=

     0     0     0
     0     0     0
     0     0     0
```

2. Controller for Design (1) - K_6

```
A=

  1.0e+03 *

  -0.1146   -0.0030   -0.0336         0   -0.0010
  -0.0136   -0.0301    0.0069    0.0000    0.0005
  -0.0347    0.0015   -0.3703    0.0010         0
  -0.0712    0.0269   -3.5465   -0.0145   -0.0029
   0.3405    0.0188   -1.0749   -0.0019   -0.0056

B=

  1.0e+03 *

   0.1146    0.0030    0.0347
   0.0119    0.0180   -0.0061
   0.0347   -0.0015    0.3703
   0.1034   -0.0247    3.4411
  -0.3191   -0.0182    1.0501

C=

 -14.4891   -0.5998   14.6035    1.7076    3.5333
  -0.0356   11.9282    0.8525    0.1936   -0.0152
 -19.1328   -0.3213  -24.5604   -3.6594    8.2734

D=

     0     0     0
     0     0     0
     0     0     0
```

3. Controller for Design (2) - K_7

```
A=

Columns 1 through 7

-181.9097   106.9798    11.9018   191.6454   720.5882    32.2075   315.5145
  33.1798  -148.0843    19.9716  -450.1834  -237.7881     5.0502   117.0712
   0.2242     0.6350   -29.5511    -0.7973    -0.6308    -0.0031     0.1117
  -0.4844     3.6851    -0.3047    -8.3162    -3.2156     0.2453     0.6790
  -1.8054     1.9433    -0.5018    -3.6390    -5.5760    -0.2499    -0.3176
  -0.0382     0.0601    -0.2716    -0.1850    -0.2105    -0.1062    -0.0028
        0          0          0          0          0          0          0
        0          0          0          0          0          0          0
        0          0          0          0          0          0          0

Columns 8 through 9

  12.5872   381.8756
   9.6405  -210.5051
  -8.4950     0.0588
  -0.0324    -0.8527
  -0.0421    -0.8143
  -0.0001    -0.0007
        0          0
        0          0
        0          0

B=

 1.0e+03 *

   7.5723     0.1510     9.1650
   2.8097     0.1157    -5.0521
   0.0027    -0.1019     0.0014
   0.0163    -0.0004    -0.0205
  -0.0076    -0.0005    -0.0195
  -0.0001     0.0000     0.0000
   0.0024          0          0
        0     0.0012          0
        0          0     0.0024

C=

Columns 1 through 7

   0.1858    -1.2947     0.1808     0.7239     0.3055     0.0011          0
   0.0364    -0.0390    -2.0583     0.0394    -0.0136    -0.0019          0
  -0.8124     0.1899     0.0158     0.0135    -0.4994    -0.0010          0

Columns 8 through 9

        0          0
        0          0
        0          0

D=

    0     0     0
    0     0     0
    0     0     0
```

REFERENCES

Al-Saggaf, U., and Franklin, G., (1988) 'Model reduction via balanced realizations: an extension and frequency weighting techniques', *IEEE Trans. Auto. Control*, 33:687-691.

Anderson, B. and Liu, Y. (1987) 'Controller reduction: concepts and approaches', *Plenary address, Proceedings of the American Control Conference, Minneapolis.*

Athans, M. (1986) 'A tutorial on the LQG/LTR method', *Proceedings of the American Control Conference, Seattle.*

Ball, J., and Ran, A. (1986) 'Hankel Norm Approximation of a rational matrix function in terms of its realization', in *Modelling, Identification and Robust Control,* (C.I. Byrnes and A. Lindquist, Eds.), North-Holland.

Bode, H. (1945) 'Network Analysis & Feedback Amplifier Design', *Van Nostrand, Princeton.*

Bucy, R. (1972) 'The Riccati equation and its bounds', *J. Comput. Systems Sci.*, 6:343-353.

Chen, M. and Desoer,C. (1982) 'Necessary and sufficient for robust stability of linear distributed feedback systems', *Int. J. Control*, 35:255-267.

Doyle, J. (1984) 'Lecture notes', *ONR/Honeywell Workshop on Advances in Multivariable Control, Honeywell Systems and Research Centre.*

Doyle, J. and Chu, C. (1986) 'Robust control of multivariable and large scale systems', *Final Technical Report, Honeywell Systems and Research Centre.*

Doyle, J., and Stein, G. (1979) 'Robustness with observers', *IEEE Trans. Auto. Control*, 24.4:607-611.

Doyle, J., and Stein, G. (1981) 'Multivariable feedback design: concepts for a classical/modern synthesis', *IEEE Trans. Auto. Control*, 26.1:4-16.

Doyle, J.,Glover, K., Khargonekar, P, Francis, B. (1988) 'State-space solutions to standard RH_2 and RH_∞ control problems', *Proceedings of the American Control Conference, Atlanta.*

Dym, H. (1988) 'J contractive matrix functions, reproducing kernel Hilbert spaces and interpolation', *CBMS Lecture Notes (draft), American Maths. Society.*

Enns, D. (1984) 'Model reduction for control system design', *PhD. Thesis, Stanford University.*

Foo, Y. and Postlethwaite, I. (1984) 'An H_∞ -minimax approach to the design of robust control systems', *Sys. and Contr. Letters.*, 5:81-88.

Francis, B. and Doyle, J. (1987) 'Linear control theory with an H_∞ optimality criterion.', *SIAM J. of Control & Opt.*, 25:815-844.

Francis, B. (1987) 'A Course in H_∞ Control Theory', *Springer Verlag, Berlin.*

Freudenberg, J., Looze, D. (1987) 'Frequency domain properties of scalar and multivariable feedback systems', *Springer Verlag, Berlin.*

Gantmacher, F. (1959) 'The Theory of Matrices', *Chelsea, New York.*

Glover, K. (1984) 'All optimal Hankel Norm Approximations of linear multivariable systems and their L_∞-error bounds', *Int. J. Control*, 39:1115-1193.

Glover, K. (1986a) 'Robust stabilization of linear multivariable systems: relations to approximation', *Int. J. Control*, 43:741-766.

Glover, K. (1986b) 'Multiplicative approximation of linear multivariable systems with L_∞ error bounds', *Proceedings of the American Control Conference, Seattle.*

Glover, K. (1987) 'Model reduction: a tutorial on Hankel Norm methods and lower bounds on L_2 errors', *10th IFAC Congress, Munich.*

Glover, K., and Doyle, J. (1988) 'State-space formulae for all stabilizing controllers that satisfy an H_∞ norm bound and relations to risk sensitivity' *Sys. and Contr. Letters, 11:167-172* .

Glover, K., and Mustafa, D. (1989) 'Derivation of the maximum entropy H_∞ controller and a state-space formula for its entropy', *Int. J. Control, to appear.*

Grimble, M. (1988) 'Optimal H_∞ multivariable robust controllers and the relationship to LQG design problems', *Int. J. Control*, 48.1:35-58.

Hinrichsen, D., and Pritchard, A. (1986) 'Stability radii of linear systems', *Sys. and Contr. Letters*, 7:1-10.

Horowitz, I. (1963) 'Synthesis of Feedback Systems', *Academic Press.*

Hung, Y.S. and MacFarlane, A.G.J. (1982) 'Multivariable feedback: a quasi-classical approach', *Springer Verlag, Berlin.*

Jonckheere, E. and Silverman, L.M. (1983) 'A new set of invariants for linear systems - applications to reduced order compensator design', *IEEE Transactions on Automatic Control.* 28.10:953-964.

Kalman, R., (1960) 'A new approach to linear filtering and prediction problems', *Journal of Basic Eng.*, 35-45.

Kasenally, E., and Limebeer, D. (1988) 'Closed formulae for a parametric mixed sensitivity problem', *(to appear Sys. and Contr. Letters).*

Khargonekar, P., Georgiou, T., Pascoal, A. (1987) 'On the robust stabilizability of linear time-invariant plants with unstructured uncertainty', *IEEE Trans. Auto. Control*, 32.3:201-207.

Khargonekar, P., Petersen, I., Zhou,K. (1987) 'Robust stabilization of uncertain systems and H_∞ optimal control', *(submitted for publication).*

Kimura, H. (1984) 'Robust stabilization for a class of transfer functions', *IEEE Trans. Auto. Control*, 29.9:788-793.

Kwakernaak, H. and Sivan, R. (1972) 'Linear Optimal Control Systems', *Wiley.*

Kwakernaak, H. (1985) 'Minimax frequency domain performance and robustness optimization of linear feedback systems', *IEEE Trans. Auto. Control*, 994-1004.

Latham, G. and Anderson, B. (1985) 'Frequency weighted optimal Hankel-Norm Approximation of stable transfer functions', *Sys. and Contr. Letters*, 5:229-236.

Limebeer, D. and Hung, S. (1987) 'An analysis of pole-zero cancellations in H_∞ optimal control problems of the first kind', *SIAM J. Control and Optimization*, 25.6:1457-1493.

Limebeer,D.J.N., Kasenally,E., (1986) 'H_∞ optimal control of a synchronous turbo-generator', *25th Conference on Decision and Control, Athens.*

Limebeer, D. and Halikias, G. (1988) 'A controller degree bound for H_∞ -optimal controllers of the second kind', *SIAM J. Control and Optimization*, 26.3:646-677.

Liu, L. and Anderson, B. (1986) 'Controller reduction via stable factorization and balancing', *Int. J. Control*, 507-531.

McFarlane, D., Glover, K., and Noton, M. (1988) 'Robust stabilization of a flexible space-platform: an H_∞ coprime factorization approach', *'Control 88',IEE Conference, Oxford.*

Maciejowski,J. (1989) 'Multivariable Feedback Design', (*to be published by Addison-Wesley*).

Meyer, D. (1987) 'Model reduction via fractional representations', PhD Thesis, *Stanford University.*

Meyer, D.G. (1988) 'A fractional approach to model reduction', *Proceedings 1988 American Control Conference, Atlanta.*

Meyer, D. and Franklin, G. (1987) 'A connection between normalized coprime factorizations and Linear Quadratic Regulator theory', *IEEE Trans. Auto. Control*, 32:227-228.

Moore B.C. (1981) 'Principal component analysis in linear systems: controllability, observability and model reduction', *IEEE Transactions on Automatic Control*, 26:17-32.

Nett, C., Jacobsen, C.A. and Balas, M.J. (1984) 'A connection between state-space and doubly coprime fractional representations.' *IEEE Transactions on Automatic Control*, 29:831-832.

Noton,M. (1987a) 'H_∞ control design for complex space structures', *Report TP9047 British Aerospace, Space and Communications Division, Bristol.*

Noton,M. (1987b) *Private communications.*

Ober, R. and McFarlane D. (1988) 'Balanced canonical forms for minimal systems: a normalized coprime factor approach', *(to appear in Linear Algebra and its Applications).*

Opdenacker, P. and Jonckheere, E. (1988) 'Characterization of passive systems through their closed-Loop LQG characteristic values', (*to appear IEEE Trans. Auto. Control).*

Pernebo, L. and Silverman, L.M., (1982), 'Model reduction via balanced state-space representations', *IEEE Transactions on Automatic Control*, 27:282-287.

Petersen, I. (1987b) 'A stabilization algorithm for a class of uncertain linear systems', *Sys. and Contr. Letters*, 8:351-357.

Postlethwaite,I., O'Young,S., Gu,D-W. (1987) 'H_∞ control system design: a critical assessment based on industrial applications', *IFAC 10th Conference Proceedings, Munich.*

Redheffer, M. (1960) 'On a certain linear fractional transformation', *J. Maths and Physics*, 39:269-286.

Rosenbrock, H. (1972) 'The stability of multivariable systems', *IEEE Trans. Auto. Control*, 17:105-107.

Rudin, W. (1986) 'Real and Complex Analysis', *McGraw-Hill, Singapore.*

Safonov, M. and Athans, M. (1977) 'Gain and phase margins of multiloop LQG regulators', *IEEE Trans. Auto. Control*, 22.2:173-179.

Safonov,M.,Chiang,R. (1988) 'CASCD using state-space L_∞ theory - a design example', *IEEE Trans. Auto. Control*, 33.5:477-479.

Safonov, M., Jonckheere, E., Verma, M., and Limebeer, D. (1987) 'Synthesis of positive real multivariable feedback systems', *Int. J. Control*, 45.3:817-842.

Safonov, M., Chiang, R. Limebeer, D. (1987) 'Hankel model reduction without balancing - a descriptor approach', in *Proc. IEEE Conf. Decision and Contr.*, Los Angeles.

Salehi,S. (1985) 'Application of adaptive observers to the control of flexible spacecraft', 10th *IFAC Symposium, 'Automatic Control in Space', Toulouse.*

Stein, G., and Athans, M. (1987) 'The LQG/LTR procedure for multivariable feedback design', *IEEE Trans. Auto. Control*, 32.2.:105-114.

Tahk, M. and Speyer, J. (1987) 'Modelling of parameter variations and asymptotic LQG synthesis', *IEEE Trans. Auto. Control*, 32.9:793-801.

Verma, M., Helton, W. and Jonckheere, E. (1986) 'Robust stabilization of a family of plants with varying numbers of right half-plane poles', *Proceedings of the American Control Conference, Seattle, Washington*, 1827-1832.

Vidyasagar, M. (1984) 'The Graph Metric for unstable plants and Robustness estimates for feedback stability', *IEEE Trans. Auto. Control*, 29:403-417.

Vidyasagar, M. (1985) 'Control System Synthesis: A Coprime Factorization Approach.' *MIT Press.*

Vidyasagar, M. (1988) 'Normalized coprime factorizations for non strictly proper systems.' *IEEE Transactions on Automatic Control*, 33:300-301.

Vidyasagar, M. and Kimura, H. (1986) 'Robust controllers for uncertain linear multivariable systems', *Automatica*, 85-94.

Willems, J. (1971) 'The Analysis of Feedback Systems', *MIT Press.*

Willems, J. (1972) 'Dissipative dynamical systems, part II: linear systems with quadratic supply rates', *Arch. Rat. Mech. Analysis*, 45:352-393.

Yedavalli, R. (1985) 'Perturbation bounds for robust stability in linear state-space models', *Int. J. Control*, 42.6:1507-1517.

Zames,G. (1966) 'On input-output stability of time-varying nonlinear feedback systems, part 1', *IEEE Trans. Auto. Control*, 11.2:228-238.

Zames, G. (1981) 'Feedback and optimal sensitivity: model reference transformations, multiplicative seminorms and approximate inverses', *IEEE Trans. Auto. Control*, 26.2:301-320.

Lecture Notes in Control and Information Sciences

Edited by M. Thoma and A. Wyner

Vol. 97: I. Lasiecka/R. Triggiani (Eds.)
Control Problems for Systems
Described by Partial Differential Equations
and Applications
Proceedings of the IFIP-WG 7.2
Working Conference
Gainesville, Florida, February 3-6, 1986
VIII, 400 pages, 1987.

Vol. 98: A. Aloneftis
Stochastic Adaptive Control
Results and Simulation
XII, 120 pages, 1987.

Vol. 99: S. P. Bhattacharyya
Robust Stabilization Against
Structured Perturbations
IX, 172 pages, 1987.

Vol. 100: J. P. Zolésio (Editor)
Boundary Control and Boundary Variations
Proceedings of the IFIP WG 7.2 Conference
Nice, France, June 10-13, 1987
IV, 398 pages, 1988.

Vol. 101: P. E. Crouch,
A. J. van der Schaft
Variational and Hamiltonian
Control Systems
IV, 121 pages, 1987.

Vol. 102: F. Kappel, K. Kunisch,
W. Schappacher (Eds.)
Distributed Parameter Systems
Proceedings of the 3rd International Conference
Vorau, Styria, July 6–12, 1986
VII, 343 pages, 1987.

Vol. 103: P. Varaiya, A. B. Kurzhanski (Eds.)
Discrete Event Systems:
Models and Applications
IIASA Conference
Sopron, Hungary, August 3-7, 1987
IX, 282 pages, 1988.

Vol. 104: J. S. Freudenberg/D. P. Looze
Frequency Domain Properties of Scalar
and Multivariable Feedback Systems
VIII, 281 pages, 1988.

Vol. 105: Ch. I. Byrnes/A. Kurzhanski (Eds.)
Modelling and Adaptive Control
Proceedings of the IIASA Conference
Sopron, Hungary, July 1986
V, 379 pages, 1988.

Vol. 106: R. R. Mohler (Editor)
Nonlinear Time Series and
Signal Processing
V, 143 pages, 1988.

Vol. 107: Y. T. Tsay, L.-S. Shieh, St. Barnett
Structural Analysis and Design
of Multivariable Systems
An Algebraic Approach
VIII, 208 pages, 1988.

Vol. 108: K. J. Reinschke
Multivariable Control
A Graph-theoretic Approach
274 pages, 1988.

Vol. 109: M. Vukobratović/R. Stojić
Modern Aircraft Flight Control
VI, 288 pages, 1988.

Vol. 110: B. J. Daiuto, T. T. Hartley,
S. P. Chicatelli
The Hyperbolic Map and Applications
to the Linear Quadratic Regulator
VI, 114 pages, 1989

Vol. 111: A. Bensoussan, J. L. Lions (Eds.)
Analysis and Optimization
of Systems
XIV, 1175 pages, 1988.

Vol. 112: Vojislav Kecman
State-Space Models of Lumped
and Distributed Systems
IX, 280 pages, 1988

Vol. 113: M. Iri, K. Yajima (Eds.)
System Modelling and Optimization
Proceedings of the 13th IFIP Conference
Tokyo, Japan, Aug. 31 – Sept. 4, 1987
IX, 787 pages, 1988.

Vol. 114: A. Bermúdez (Editor)
Control of Partial Differential Equations
Proceedings of the IFIP WG 7.2
Working Conference
Santiago de Compostela, Spain, July 6–9, 1987
IX, 318 pages, 1989

Vol. 115: H.J. Zwart
Geometric Theory for Infinite
Dimensional Systems
VIII, 156 pages, 1989.

Vol. 116: M.D. Mesarovic, Y. Takahara
Abstract Systems Theory
VIII, 439 pages, 1989

Lecture Notes in Control and Information Sciences

Edited by M. Thoma and A. Wyner

Vol. 117: K.J. Hunt
Stochastic Optimal Control Theory
with Application in Self-Tuning Control
X, 308 pages, 1989.

Vol. 118: L. Dai
Singular Control Systems
IX, 332 pages, 1989

Vol. 119: T. Başar, P. Bernhard
Differential Games and Applications
VII, 201 pages, 1989

Vol. 120: L. Trave, A. Titli, A. M. Tarras
Large Scale Systems:
Decentralization, Structure Constraints
and Fixed Modes
XIV, 384 pages, 1989

Vol. 121: A. Blaquière (Editor)
Modeling and Control of Systems
in Engineering, Quantum Mechanics,
Economics and Biosciences
Proceedings of the Bellman Continuum
Workshop 1988, June 13–14, Sophia Antipolis, France
XXVI, 519 pages, 1989

Vol. 122: J. Descusse, M. Fliess, A. Isidori,
D. Leborgne (Eds.)
New Trends in Nonlinear Control Theory
Proceedings of an International
Conference on Nonlinear Systems,
Nantes, France, June 13–17, 1988
VIII, 528 pages, 1989

Vol. 123: C. W. de Silva, A. G. J. MacFarlane
Knowledge-Based Control with
Application to Robots
X, 196 pages, 1989

Vol. 124: A. A. Bahnasawi, M. S. Mahmoud
Control of Partially-Known
Dynamical Systems
XI, 228 pages, 1989

Vol. 125: J. Simon (Ed.)
Control of Boundaries and Stabilization
Proceedings of the IFIP WG 7.2 Conference
Clermont Ferrand, France, June 20–23, 1988
IX, 266 pages, 1989

Vol. 126: N. Christopeit, K. Helmes
M. Kohlmann (Eds.)
Stochastic Differential Systems
Proceedings of the 4th Bad Honnef Conference
June 20–24, 1988
IX, 342 pages, 1989

Vol. 127: C. Heij
Deterministic Identification
of Dynamical Systems
VI, 292 pages, 1989

Vol. 128: G. Einarsson, T. Ericson,
I. Ingemarsson, R. Johannesson,
K. Zigangirov, C.-E. Sundberg
Topics in Coding Theory
VII, 176 pages, 1989

Vol. 129: W. A. Porter, S. C. Kak (Eds.)
Advances in Communications and
Signal Processing
VI, 376 pages, 1989.

Vol. 130: W. A. Porter, S. C. Kak,
J. L. Aravena (Eds.)
Advances in Computing and Control
VI, 367 pages, 1989

Vol. 131: S. M. Joshi
Control of Large Flexible Space Structures
IX, 196 pages, 1989.

Vol. 132: W.-Y. Ng
Interactive Multi-Objective Programming
as a Framework for Computer-Aided Control
System Design
XV, 182 pages, 1989.

Vol. 133: R. P. Leland
Stochastic Models for Laser Propagation
in Atmospheric Turbulence
VII, 145 pages, 1989.

Vol. 134: X. J. Zhang
Auxiliary Signal Design in Fault
Detection and Diagnosis
XII, 213 pages, 1989

Vol. 135: H. Nijmeijer, J. M. Schumacher (Eds.)
Three Decades of Mathematical System Theory
A Collection of Surveys at the Occasion of the
50th Birthday of Jan C. Willems
VI, 562 pages, 1989

Vol. 136: J. Zabczyk (Ed.)
Stochastic Systems and Optimization
Proceedings of the 6th IFIP WG 7.1
Working Conference,
Warsaw, Poland, September 12–16, 1988
VI, 374 pages. 1989